CONTENTS

Preface 2

Bibliographical and Historiographical Studies 3

Source Collections and Reference Works 4

Periodicals 5

Attitudes towards Nature 6

American Environmental History 11

Wildlife and Animal Rights 21

Water, Soil and Minerals 23

Parks and Public Land 26

Forests 29

Regional Studies 32

Biographical Studies 35

Environmental Ethics 37

Comparative Studies (Selected) 40

A Chronology of Major Legislation and Events
in American Environmental History 43

**Other guides to the British Library's North American Collections
published by the Eccles Centre**

An Era of Change: Contemporary US-UK-West European Relations

American Slavery: Pre-1866 Imprints

Mormon Americana

United States Government Policies Toward Native Americans, 1787-1990

United States and Canadian Newspaper Holdings in the
British Library Newspaper Library

Mining the American West

Imagining the West

The Harlem Renaissance

PREFACE

The image of America as a land of unlimited resources exerted a strong influence on her explorers and settlers. It has been argued that this great economic abundance has helped shape the American character. The harvesting of the rich soils, forest, animal and mineral wealth have surely given Americans a lifestyle envied the world over. But these have also helped to create a sense of wastefulness, materialism, gluttony, and environmental pollution. And some argue that their exploration has fostered economic and social inequality.

One of the fascinating aspects of American history is the emergence of an environmental movement. From the game and forest cutting restrictions of colonial America, to the growing environmental conscientiousness in the nineteenth century to the growing institutional responses to these problems in the twentieth century, the story is as filled with significance for the present and the future.

The purpose of this bibliography is to list the significant studies of American Environment History found in the British Library. Because no clear line can be drawn between conservation, reclamation, ecology and natural history, each is included. This bibliography does not contain references to all the holdings on the topic in the British Library, but the serious student will find the most significant studies herein. And because environmental issues are international in scope, a comparative section is included as well. Finally a chronology of events and governmental legislation is provided.

David J. Whittaker

BIBLIOGRAPHICAL AND HISTORIOGRAPHICAL STUDIES

ANGLEMYER, Mary and Eleanor R. Seagraves.
The Natural Environment: An Annotated Bibliography of Attitudes and Values (Washington, D.C.: Smithsonian Institution Press, 1984). [lists over 800 items] (2725.e.264)

DODDS, Gordon B.
"Conservation and Reclamation in the Trans-Mississippi West: A Critical Bibliography," *Arizona and the West* 13 (1971):143-171. (P.701/1302)
"The Historiography of American Conservation: Past and Prospects," *Pacific Northwest Quarterly* 56 (April 1965):75-81. (P.P.8004.jv)

DURRENBERGER, Robert W.
Environment and Man: A Bibliography (Palo Alto, CA: National Press, 1970)

FAHL, Ronald J.
North American Forest and Conservation History: A Bibliography (Santa Barbara, CA: ABC-Clio Press, 1976).

HURT, R. Douglas, and Mary Ellen Hurt.
The History of Agricultural Science and Technology: An International Annotated Bibliography (New York: Garland Publishing Co., 1994). (2725.e.2379)

KELLERT, Stephen R. and Joyce K. Berry.
A Bibliography of Human/Animal Relations (Lanham, MD: University Press of America, 1985). [4,000 items organised into 50 topical areas] (2725.d.126)

LANG, William L.
"Using and Abusing Abundance: The Western Resource Economy and the Environment," in *Historians and the American West,* Michael P. Malone ed., (Lincoln, NE: University of Nebraska Press, 1983), pp. 270-99. (X.800/37290)

LEDUC, Thomas.
"The Historiography of Conservation," *Journal of Forest History* 9 (October 1965):23-28.

LEE, Lawrence B.
"100 Years of Reclamation Historiography," *Pacific Historical Review* 37 (November 1978):507-64. (Ac.8504.c)
Reclaiming the American West: An Historiography and Guide (Santa Barbara, CA: ABC-Clio, 1980). (X.800/29040)

NASH, Roderick.
"American Environmental History: A New Teaching Frontier," *Pacific Historical Review* 41 (August 1972):362-72. (Ac.8504.c)
"The State of Environmental History," in *The State of American History,* Herbert J. Bass, ed. (Chicago, IL: Quadrangle Books, 1970), pp. 249-60. (X.800/7977)

OPIE, John.
"The Environment and the Frontier," in *American Frontier and Western Issues: A Historiographical Review,* Roger L. Nichols, ed. (Westport, CT: Greenwood Press, 1986), pp.7-25.
"Environmental History in the West," in *The Twentieth-Century West, Historical Interpretations,* Gerald D. Nash and Richard W. Etulain, eds. (Albuquerque, NM: University of New Mexico Press, 1989), pp. 309-232. (YA.1990.b.150)
"Environmental History: Pitfalls and Opportunities," *Environmental Review* 7 (Winter 1983):8-16. (Boston Spa)
"Frontier History in Environmental

Perspective," in *The American West: New Perspectives, New Dimensions,* Jerome O. Steffen, ed. (Norman, OK: University of Oklahoma Press, 1979), pp. 9-34.

RAKESTRAW, Lawrence.
"Conservation Historiography: An Assessment," *Pacific Historical Review* 41 (August 1972):271-88. (Ac.8504.c)

SCHLEBECKER, John T., ed.
Bibliography of Books and Pamphlets on the History of Agriculture in the United States, 1607-1967 (Santa Barbara, CA: ABC-Clio, 1969). (2729.a.55)

SESSIONS, George
"The Deep Ecology Movement: A Review," *Environmental Review* 11 (1987):105-25. (Boston Spa)
"Shallow and Deep Ecology: A Review of the Philosophical Literature, " in *Ecological Consciousness: Essays from the Earthday X Colloquium, University of Denver, April 21-24, 1980,* Robert C. Schultz and J. Donald Hughes, eds. (Washington, D.C., 1981), pp. 391-462.

TERRIE, Philip G.
"Recent Work in Environmental History," *American Studies International* 27 (October 1989):42-65.

WHITE, Richard.
"American Environmental History: The Development of a New Historical Field," *Pacific Historical Review* 54 (August 1985):297-335 (Ac.8504.c)

WORSTER, Donald, ed.
"History as Natural History: An Essay on Theory and Method," *Pacific Historical Review* 53 (February 1984):1-19. (Ac.8504.c)

SOURCE COLLECTIONS AND REFERENCE WORKS

DAVIS, RICHARD C.
Encyclopedia of American Forest and Conservation History, 2 volumes (New York: Macmillan Publishing Co., 1983).
"Environmental History," *Pacific Historical Review* 41 (August 1972):271-372. [Special issue devoted to the topic] (Ac.8504.c.)

JARRETT, Henry, ed.
Perspectives on Conservation: Essays on America's Natural Resources (Baltimore, MD: Johns Hopkins Press, 1958). (08232.g.7)

MCHENRY, Robert and Charles Van Doren, eds.
A Documentary History of Conservation in America (New York: Praeger, 1972). (X.320/3387)

NATIONAL WILDLIFE FEDERATION
Conservation Directory (Washington, D.C.: NWF, annual edition).

NEIDERHEISER, Clodough M.
Forest History Sources of the United States and Canada: A Compilation of the Manuscript Sources of Forestry, Forest Industry, and Conservation History (St Paul, MN, 1956).

OWINGS, Loren C.
Environmental Values, 1860-1972: A Guide to Information Sources (Detroit, MI: Gale Research, 1976).

"A Roundtable: Environmental History," *Journal of American History* 76 (March 1990):1087-1147. [Includes William Cronon, Alfred Crosby, Carolyn Merchant, Stephen Pyne, Richard White and Donald Worster] (Ac.8408/2)

SMITH, Frank E., ed.
Conservation in the United States: A Documentary History 5 Volumes (New York: Chelsea House, 1971-). (X.0320/67)

STOKES, Samuel N. with A. Elizabeth Watson, Genevieve P. Keller and J, Timothy Keller.
Saving America's Countryside: A Guide to Rural Conservation (Baltimore MD: Johns Hopkins University Press for the National Trust for Historic Preservation, 1989). [Includes an annotated bibliography and an appendix listing names and addresses of non-profit organizations involved in conservation issues/concerns plus a list of Federal Agencies involved in the same] (YC,1989.b.7982)

STROUD, Richard H., ed.
National Leaders of American Conservation (Washington, D.C.: Smithsonian Institution Press, 1985).

WILD, Peter.
Pioneer Conservationists of Eastern America (Missoula, MT: Mountain Press Publishing Co., 1986). (YA.1988.b.5237)
  Pioneer Conservationists of Western America (Missoula MT: Mountain Press Publishing Co., 1979). (X.520/36883)

PERIODICALS

Environmental Review [1974-] [Journal of the American Society for Environmental History] (Boston Spa)

Environmental Ethics: An Interdisciplinary Journal Dedicated to the Philosophical Aspects of Environmental Problems [1979-] (P.521/2330)

ECOLOGY, A Quarterly Journal Devoted to all Phases of Ecological Biology [published by the Ecological Society of America] [from volume 34, 1953] (Ac.3313/2)
Journal of Forest History

Ethics and Animals

ATTITUDES TOWARD NATURE

ABBEY, Edward.
Desert Solitaire: A Season in the Wilderness (London: Robin Clark, 1992). (YK.1992.a.4939)
The Journey Home: Some Words in Defense of the American West (New York: Dutton, 1977).
The Monkey Wrench Gang (Edinburgh: Canongate Publishing, 1978). (Nov. 23999)

ANDERSON, David D., ed.
Sunshine and Smoke: American Writers and The American Environment (Philadelphia, PA: J.B. Lippincott, 1971).

AUSTIN, Mary Hunter.
The Land of Little Rain [a novel] (Boston, MA: Houghton Mifflin, 1903). (10409.f.35)

BERGON, Frank, ed.,
The Wilderness Reader (New York: Mentor, 1980).

BERRY, Wendell.
Home Economics: Fourteen Essays (San Francisco, CA: North Point Press, 1987). (YA.1989.a.20149.)
The Unsettling of America: Culture and Agriculture (San Francisco, CA: Sierra Club, 1978). (X.520/21689)
Standing on Earth: Selected Essays (Ipswich: Golgonooza, 1991). (YC.1991.a.1399)

BROOKS, Paul.
The House of Life: Rachel Carson at Work, with Selections from her Writings, Published and Unpublished (Boston, MA: Houghton Mifflin, 1972). (X.320/4121)
Speaking for Nature: How Literary Naturalists from Henry Thoreau to Rachel Carson Have Shaped America (Boston, MA: Houghton Mifflin, 1980).

BROWER, David R, ed.
The Meaning of Wilderness to Science (San Francisco, CA: Sierra Club, 1960). (X.311/1263)
Wilderness: America's Living Heritage (San Francisco, CA: Sierra Club, 1961). (X.311/1255)
Wildlands in Our Civilization (San Francisco, CA: Sierra Club, 1964). (X.311/1251)

BURROUGHS, John.
John Burroughs' America. Selections from the Writings of the Hudson River Naturalist, Farida A. Wiley ed. (New York: Devin-Adair Co., 1952). (W.P. 13887/3)
Writings of John Burroughs, 20 volumes (London: J.M. Dent and Co., 1895-1919). (012295.a.6)

CALLENBACH, Ernest.
Ectopia: A Novel about Ecology, People and Politics in 1999 (1975; London: Pluto Press, 1978). (Nov.37480)

CALLICOTT, J. Baird, ed.
Companion to 'A Sand Country Almanac': Interpretive and Cultural Essays (Madison, WI: University of Wisconsin Press, 1987).

CARSON, Rachel L.
The Sea Around Us (New York: Oxford University Press, 1951). (10001.ee.46) [2nd ed., London, 1955 (10498.tt.33)]
Silent Spring (London: Hamish Hamilton, 1963). (N.R.L.S.I. (B.)) [1982 Penguin edition (X.529/50495)]

CHASE, Steve, ed.
Defending the Earth: A Dialogue Between Murray Bookchin and Dave Foreman (Boston, MA: South End Press, 1991). [Foreman was a founder of Earth First!] (YA.1993.a.4886)

CLOUGH, Wilson.
The Necessary Earth: Nature and Solitude in American Literature (Austin, TX: University of Texas Press, 1964). (X.909/8301)

DANSEREAU, Pierre.

Inscape and Landscape: The Human Perception of Environment (New York: Columbia University Press, 1975).

DILLARD, Annie.
An American Childhood (London: Picador, 1988). (YK.1991.a.5440)
Pilgrim at Tinker Creek (London: Pan Books, 1976). (X.319/16701)
Teaching a Stone to Talk: Expeditions and Encounters (London: Pan Books, 1984). (X.958/26533)

EKIRCH, Arthur A.
Man and Nature in America (New York: Columbia University Press, 1963). (08203.f.16)

FLEXNER, Janes Thomas.
That Wilder Image: The Painting of America's Native School from Thomas Cole to Winslow Homer (Boston, MA: Little Brown, 1962). (X.421/1205)

FOESTER, Norman.
Nature in American Literature: Studies in the Modern View of Nature (New York: Macmillan, 1923). Reprinted New York: Russell and Russell, 1950.

FOREMAN, Dave and Bill Haywood, eds.
Ecodefence: A Field Guide to Monkeywrenching (Tucson, AZ: Ned Ludd, 1987).

FOX, W.
"The Deep Ecology-Ecofeminism Debate and its Parallels," *Environmental Ethics* 11 (1989):5-25. (P.521/2330)

FUSSELL, Edwin.
Frontier: American Literature and the American West (Princeton, NJ: Princeton University Press, 1966). (X.909/20286)

GLACKEN, Clarence J.
Traces on the Rhodian Shore: Nature and Culture in Western Thought from Ancient Times to the End of the Eighteenth Century (1967; Berkeley, CA: University of California Press, 1976). (X.529/40942)

GOETZMANN, William H. and William N. Goetzmann.
The West of the Imagination (New York: Norton, 1986). (YV.1988.b.912)

GRABNER, Linda H.
Wilderness as Sacred Space (Washington, D.C.: Association of American Geographers, 1976).

HADLEY, Edith Jane.
"John Muir's Views of Nature and Their Consequences," (Ph.D. dissertation, University of Wisconsin, Madison, 1956).

HUTH, Hans.
Nature and the American: Three Centuries of Changing Attitudes (Berkeley, CA: University of California Press, 1957). (10414.dd.8)

JACKSON, John Brinckerhoff.
Discovering the Vernacular Landscape (New Haven, CT: Yale University Press, 1984). (X.322/14995)

KAY, Jeanne and Craig J. Brown.
"Mormon Beliefs about Land and Natural Resources, 1847-1877," *Journal of Historical Geography* 11 (July 1985):253-267. (P.801/3025)

KLINE, Marcia B.
Beyond the Land Itself: Views of Nature in Canada and the United States (Cambridge, MA: Harvard University Press, 1970).

KOLODNY, Annette.
The Land Before Her: Fantasy and Experience of the American Frontiers, 1630-1860 (Chapel Hill, NC: University of North Carolina

Press, 1984). (YH.1988.a.882)
The Lay of the Land: Metaphor as Experience and History in American Letters (Chapel Hill, NC: University of North Carolina Press, 1975). (X.981.21568)

KRUTCH, Joseph W.
The Best Nature Writing of Joseph Wood Krutch (New York: William Morrow and Co., 1969). (X.329/5552)
"Conservation is Not Enough," *The American Scholar* 23 (1954):295-305. (P.P. 6365.bg)
The Voice of the Desert, A Naturalist's Interpretation (London: Alvin Redman, 1956). (7009.r.2)

LEISS, William.
The Domination of Nature (New York: George Braziller, 1972). (X.329/7290)

LEOPOLD, Aldo.
A Sand Country Almanac and Sketches Here and There (New York: Oxford University Press, 1949). (7008.e.18)

LOPEZ, Barry.
Arctic Dreams: Imagination and Desire in a Northern Landscape (London: Macmillan, 1986). (YC.1986.b.3590)
Crossing Open Ground (London: Macmillan, 1988). (YC.1988.b.5972)
Desert Notes and River Notes (London: Picador, 1990). (YC.1991.a.885)
Of Wolves and Men (London: Dent, 1978). (X.320/11732)
The Rediscovery of North America (Lexington, KY: University Press of Kentucky, 1990). (YA.1993.a.21032)

LYON, Thomas J.
"The Nature Essay in the West", in *A Literary History of the American West*, eds. J. Golden Taylor and Thomas J. Lyon (Fort Worth, TX: Texas Christian University Press, 1987), pp.221-265. (YH.1987.b.595)

MCKIBBEN, Bill.
The End of Nature (London: Viking, 1990). (YK. 1990.a.2883)

MCPHEE, John.
Encounters with the Archdruid (Narratives About a Conservationist [David R. Brower] and Three of His Natural Enemies) (New York: Farrar Straus and Giroux, 1971).
The Control of Nature (London: Hutchinson Radius, 1990). (YC.1990.b.5905)
Basin and Range (New York: Farrar Straus Girox, 1981). (X.950/20201)
Coming into the Country (London: H.Hamilton, 1978). (X.809/43658)

MARSH, George Perkins.
Man and Nature: or, Physical Geography as Modified by Human Action (London, 1864). (1006.dd.1) [1874 London ed: *The Earth as Modified by Human Action* (7001.ee.19); 1965 Harvard University Press edition, edited by David Lowenthal (11485.b.1/44)]

MARSHALL, Peter H.
Nature's Web: An Exploration of Ecological Thinking (London: Simon and Schuster, 1992). (YK.1993.b.7327)

MARX, Leo.
The Machine in the Garden, Technology and the Pastoral Idea in America (Oxford: Oxford University Press, 1964(. (X.519/963)

MILLER, Charles A.
Jefferson and Nature: An Interpretation (Baltimore, MD: Johns Hopkins University Press, 1988). (YH.1989.b.594)

MITCHELL, Lee Clark.
Witness to a Vanishing America: The Nineteenth Century Response (Princeton, NJ: Princeton

University Press, 1981). (X.800/31071)

MUIR, John.
The Mountains of California (New York: Century Co., 1894). (10411.df.31)
   My First Summer in the Sierra.... (London: Constable & Co., 1911). (10410.t.17)
   Our National Parks (Boston, MA: Houghton Mifflin, 1901). (010409.ff.20)
   The Yosemite (New York: Century Co., 1912). (10410.pp.16)
   The Life and Letters of John Muir, 2 Volumes, William F. Bade, ed. (Boston, MA: Houghton Mifflin 1923-1924). (010884.df.28)
   Travels in Alaska (Boston, MA: Houghlin Mifflin, 1915). (10470.r.13)
   Works, The Sierra Edition, 10 Volumes, William Frederick Bade, ed. (Boston, MA: Houghton Mifflin, 1915-1924).

MUMFORD, Lewis.
The Myth of the Machine, Technics and Human Development (London: Secker and Warburg, 1967). (X.800/2490)

MURPHY, Earl Finbar.
Governing Nature (Chicago, IL: Quadrangle Books, 1967). (X.329/4499)
   Man and His Environment: Law (New York: Harper and Row, 1971). (X.200/4920)
   Water Purity: A Study in Legal Control of Natural Resources (Madison, WI: University of Wisconsin Press, 1961).

NASH, Roderick.
Wilderness and the American Mind, 3rd ed. Revised (New Haven, CT: Yale University Press, 1982). (X.529/54571)

ODELL, Rice.
Environmental Awakening: The New Revolution to Protect the Earth (Cambridge, MA: Ballinger, 1980). (X.322/13160)

OELSCHLAEGER, Max.
The Idea of Wilderness: From Prehistory to the Age of Ecology (New Haven, CT: Yale University Press, 1991). (YC.1991.b.5950)
   ed. *The Wilderness Condition: Essays on Environment and Civilization* (San Francisco, CA: Sierra Club Books, 1992). (YA.1992.a.21030)

OPHULS, William, and A. Stephen Boyan.
Ecology and the Politics of Scarcity Revisited: The Unraveling of the American Dream (New York: W.H. Freeman, 1992). (YC.1993.b.3495)

POTTER, David M.
People of Plenty, Economic Abundance and the American Character (Chicago, IL: University of Chicago Press, 1954). (Ac.2691.dw(26))

SANFORD, Charles L.
The Quest for Paradise: Europe and the American Moral Imagination (Urbana, IL: University of Illinois Press, 1961). (08413.bb.8)

SCHMITT, Peter.
Back to Nature: The Arcadian Myth in Urban America (New York: Oxford University Press, 1969). (X.809/7535)

SHEPARD, Paul.
Man in the Landscape: A Historic View of the Esthetics of Nature (1967; College Station, TX: Texas A & M University Press, 1991). (YA.1993.a.19581)

SLOTKIN, Richard.
Regeneration through Violence: The Mythology of the American Frontier, 1600-1860 (Middletown, CT: Wesleyan University Press, 1973).
   The Fatal Environment: The Myth of the Frontier in the Age of Industrialization, 1800-1890 (Middletown, CT: Wesleyan University Press, 1986). (YC.1987.a.5907)
   Gunfighter Nation: The Myth of the Frontier

in Twentieth-Century America (New York: Atheneum, 1992). (YA.1993.b.6151)

SLOVIC, Scott.
Seeking Awareness in American Nature Writing: Henry Thoreau, Annie Dillard, Edward Abbey, Wendell Berry, Barry Lopez (Salt Lake City, UT: University of Utah Press, 1992). (YA.1992.a.22932)

SMALLWOOD, William Martin.
Natural History and the American Mind (New York: Columbia University Press, 1941). (Ac.2688/49)

THOMAS, Keith.
Man and the Natural World: Changing Attitudes in England 1500-1800 (London: Allen Lane, 1983). (X.520/30513)

THOREAU, Henry David.
A Week on the Concord and Merrimack Rivers (Boston, MA: 1849). (10411.f.22)
　　Walden; or Life in the Woods (Boston, MA: 1854). (12356.b.28)
　　The Maine Woods (Boston, MA: 1864). (10411.aaa.41)

TICHI, Cecelia.
New World, New Earth: Environmental Reform in American Literature from the Puritans through Whitman (New Haven, CT: Yale University Press, 1979). (X.981/21963)

TOKAN, Brian.
The Green Alternative: Creating an Ecological Future (San Pedro, CA: R&E Miles, 1987). (YC.1991.a.4788)

TUAN, Yi-Fu.
Passing Strange and Wonderful: Aesthetics, Nature and Culture (Washington, D.C.: Island Press, 1993).
　　Space and Place: The Perspective of Experience (London: Edward Arnold, 1977). (X.421/10526)
　　Topophilia: A Study of Environmental Perception, Attitudes, and Values (Englewood Cliffs, NJ: Prentice Hall, 1974). (X.519/19857)

TUCKER, William.
Progress and Privilege (Garden City, NY: Anchor/Doubleday, 1982).

TURNER, Frederick Jackson.
The Frontier in American History (New York: H. Holt and Co., 1921). (9616.aa.8)

WILLIAMS, George H.
Wilderness and Paradise in Christian Thought (New York: Harper, 1962).

WILLIAMS, Terry Tempest.
Refuge: An Unnatural History of Family and Place (New York: Pantheon Books, 1991).

WILSON, Alexander.
The Culture of Nature, North American Landscape from Disney to the Exxon Valdez (Oxford: Blackwell, 1992). (YK.1992.a.11580)

AMERICAN ENVIRONMENTAL HISTORY

ALBRECHT, Stan L.
"Legacy of the Environmental Movement," *Environment and Behavior* 8 (1976):147-68. (P.521/3445)

ALLIN, Craig W.
The Politics of Wilderness Preservation (Westport, CT: Greenwood Press, 1982). (X.322/11424)

BALDWIN, Donald N.
The Quiet Revolution: The Grass Roots of Today's Wilderness Preservation Movement (Boulder, Co: Pruett Publishing Co., 1972).

BALL, Howard.
Justice Downwind: America's Atomic Testing Program in the 1950s (New York: Oxford University Press, 1986). (YH.1987.b.198)

BATES, J. Leonard.
"Fulfilling American Democracy: The Conservation Movement, 1907 to 1921," *Mississippi Valley Historical Review* 44 (June 1957):29-57. (Ac.8408/2)
  "The Midwest Decision, 1915: A Landmark in Conservation History," *Pacific Northwest Quarterly* 51 (January 1960):26-34. (P.P. 8004.jv)
  The Origins of Teapot Dome: Progressives, Parties and Petroleum, 1909-1921 (Urbana, IL: University of Illinois Press, 1963). (8184.a.24)

BELASCO, Warren James.
Americans on the Road: From Autocamp to Motel, 1910-1945 (Cambridge, MA: MIT Press, 1979). (X.800/29010)

BERGER, John J.
Restoring the Earth: How Americans are Working to Renew Our Damaged Environment (New York: Knopf, 1985).

BERLANGER, Dian Olson.
Managing American Wildlife: A History of the International Association of Fish and Wildlife Agencies (Amherst, MA: University of Massachusetts, 1988).

BOWERS, William L.
The Country-Life Movement in America, 1900-1920 (Port Washington, NY: Kennikat, 1974).

CALDWELL, Lynton K.
Between Two Worlds: Science, the Environmental Movement, and Public Choice (Cambridge: Cambridge University Press, 1990). (YC.1990.b.5608)
  Biocracy: Public Policy and the Life Sciences (Boulder, CO: Westview, 1987). (YC.1988.b.5448)
  International Environmental Policy: Emergence and Dimensions (Durham, NC: Duke University Press, 1984). (YA.1988.b.5242)
  Science and the National Environmental Policy Act: Redirecting Policy through Procedural Reform (Tuscaloosa, AL: University of Alabama Press, 1982).
  U.S. Interests and the Global Environment, Occasional Papers, Stanley Foundation (Muscatine, IA: Stanley Foundation, 1985). (X.529/71707)

CALDWELL, Lynton K., Lynton R. Hayes and Isabel M. MacWirter.
Citizens and the Environment: Cases Studies in Popular Action (Bloomington, IN: Indiana University Press, 1976). (X.320/10634)

CALLICOTT, J. Baird, ed.
"American Indian Land Wisdom? Sorting Out the Issues," *Journal of Forest History* 33 (1989):35-42.

CARROLL, Peter N.
Puritanism and the Wilderness: The Intellectual Signficance of the New England Frontier, 1629-

1700 (New York: Columbia University Press, 1969). (X.100/7552)

CHASE, Stuart.
Rich Land, Poor Land: A Study of Waste in the Natural Resources of America (New York: McGraw-Hill, 1936). (10409.y.24)

CLARK, J. Stanley.
The Oil Century: From Drake Well to the Conservation Era (Norman, OK: University of Oklahoma Press, 1958). (7114.aa.10)

CLEMENTS, Kendrick.
"Herbert Hoover and Conservation, 1921-33," *American Historical Review* 84 (February 1984):67-88. (P.P. 3437. baa)

CLEPPER, Henry. ed.
Careers in Conservation: Opportunities in Natural Resources Management (New York: Chechester Wiley for the Natural Resources Council of America, 1979). (X.520/14242)
Origins of American Conservation (New York: Ronald Press, 1966).

COATES, Peter.
In Nature's Defense: Conservation and Americans, British Association for American Studies, Pamphlet No. 26 (Keele, UK: Ryburn Productions, Keele University, 1993).

COHEN, Michael P.
The History of the Sierra Club, 1892-1970 (San Francisco, CA: Sierra Club Books, 1988). (YA.1989.b.6602)

COOLEY, Richard A. and Geoffrey Wandesforde-Smith, eds.,
Congress and the Environment (Seattle, WA: University of Washington Press, 1970). (X.800/4592)

CRONON, William.
Changes in the Land: Indians, Colonists and the Ecology of New England (New York: Hill and Wang, 1983). (X.529/69451)
"Landscapes of Abundance and Scarcity," in *The Oxford History of the American West*, Clyde A. Milner II, Carol A. O'Connor and Martha A. Sandweiss, eds. (New York: Oxford Univesity Press, 1994), pp. 605-637.
Nature's Metropolis, Chicago and the Great West (New York: Norton, 1991). (YC.1992.b.2740)

CROSBY, Alfred W.
The Columbian Exchange: Biological and Cultural Consequences of 1492 (Westport, CT: Greenwood Press, 1972).
Ecological Imperialism, The Biological Expansion of Europe, 900-1900 (Cambridge: Cambridge University Press, 1986). (YK.1986.b.602)
"Virgin Soil Epidemics as a Factor in the Aboriginal Depopulation in America," *William and Mary Quarterly*, 3rd Series, 33 (1976):289-99. (Ac.8543)

COYLE, David Cushman.
Conservation: An American Story of Conflict and Accomplishment (New Brunswick, NJ: Rutgers University Press, 1957). (8221.c.72)

CROSS, Whitney R.
"Ideas in Politics: The Conservation Politics of the Two Roosevelts," *Journal of the History of Ideas* 14 (1953):421-38. (W.P.1159)
"W. J. McGee and the Idea of Conservation," *Historian* 15 (1953):148-62.

DASMANN, Raymond F.
Environmental Conservation (New York: John Wiley and Sons, 2nd ed. 1968). (X.311/2295)
The Last Horizon (New York: Macmillan, 1963). (7012.d.17)

DAVIS, John, ed.
The Earth First! Reader: Ten Years of Radical Environmentalism (Salt Lake City, UT: Peregrine Smith, 1991). (YA.1992.a.10520)

DAVISON, Art.
In the Wake of the Exxon Valdez: The Devastating Impact of the Alaska Oil Spill (San Francisco, CA: Sierra Club Books, 1990).

DODDS, Gordon B.
"The Stream-Flow Controversy: A Conservation Turning Point," *Journal of American History* 56 (June 1969):56-69. (Ac.8408/2)

DOUGHTY, Robin W.
The English Sparrow in the American Landscape: A Paradox in Nineteenth Century Wildlife Conservation (Oxford: Oxford Publishing Co. for the School of Geography, University of Oxford, 1978).   (P.805/202)
Feather Fashions and Bird Preservation: A Study in Nature Protection (Berkeley, CA: University of California Press, 1975). (X.320/10070)

DUKE, Frederick, William L. Howenstine, and June Sochen, eds.
Destroy to Create: Interaction with the Natural Environment in the Building of America (Hinsdale, IL: Dryden, 1972).

DUNLOP, Riley E.
"Public Opinion and Environmental Policy," in *Environmental Politics and Policy, Theories and Evidence*, James P. Lester, ed. (Durham, NC: Duke University Press, 1989), pp. 87-134. (YC.1993.b.2128)
"Trends in Public Opinion Toward Environmental Issues: 1965-1990," in *American Environmentalism: The U.S. Environmental Movement, 1970-1990*, Riley E. Dunlay and Angela G. Mertig, eds. (Philadelphia, PA: Taylor and Francis, 1992), pp. 89-116. (YC.1993.b.1277)

DUNLOP, Riley E. and Angela G. Mertig, eds.
American Environmentalism: The U.S. Environmental Movement, 1970-1980 (London: Taylor and Francis, 1992). (YC.1993.b.1277)

DUNLOP, Thomas.
DDT: Scientists, Citizens and Public Policy (Princeton, NJ: Princeton University Press, 1981). (X.629/15972)
"Sport Hunting and Conservation, 1880-1920," *Environmental Review* 12 (Spring1988): 51-100. (Boston Spa)
"Values for Varmints: Predator Control and Environmental Ideas, 1920-1939," *Pacific Historical Review* 53 (May 1984):141-61. (Ac.8504.c)

DUPREE, A. Hunter.
Science in the Federal Government: A History of Policies and Activities to 1940 (Cambridge, MA: Belknap Press of Harvard University Press, 1957). (9617.n.1)

EASTON, Robert.
Black Tide [A history of the oil spill near Santa Barbara, CA in 1969] (New York: Delacorte, 1972).

The Environmental Committees: A History of the House, Senate Interior, Agriculture and Science Committees Ralph Nader Congress Projects (New York: Grossman, 1975). (X.529/37064)

FAIRHALL, David and Philip Jordan.
Black Tide Rising: The Wreck of the 'Amoco Cadiz' (London: Deutsch, 1980). (X.800/28947)

FLEMING, Donald.
"Roots of the New Conservation Movement," *Perspectives in American History* 6 (1972):7-91. (ZC.9.a.1583)

FLINK, James J.
The Automobile Age (Cambridge, MA: MIT Press, 1988). (YH.1988.b.1179)

FLORES, Dan L.
"Agriculture, Mountain Ecology, and the Land Ethic: Phases of the Environmental History of Utah," in *Working the Range: Essays on the History of Western Land Management and the Environment,* John R. Wunder, ed. (Westport, CT: Greenwood Press, 1985), pp. 157-86. (YC.1988.a.4053)
  "Bison Ecology and Bison Diplomacy: The Southern Plains from 1825 to 1850," *Journal of American History* 78 (September 1991):465-85. (Ac.8408/2)

FOX, Stephen.
John Muir and His Legacy: The American Conservation Movement (Boston, MA: Little, Brown, 1981). (X.322/14485)

FRADKIN, Philip L.
Fallout: An American Tragedy (Tucson, AZ: University of Arizona Press, 1989). (YC.1989.a.8404)

FRANCIS, John G.
"Environmental Values, Intergovernmental Policies, and the Sagebrush Rebellion," in *Western Public Lands: The Management of Natural Resources in a Time of Declining Federalism,* John G. Francis and Richard Ganzel, eds. (Totowa, NJ: Rowman and Allenheld, 1984), pp. 29-45. (YA.1987.b.2631)

FRIENDS OF THE EARTH, et al.
Ronald Reagan and the American Environment: An Indictment (Andover, MA: Brick House, 1982). (YA.1988.b.3948)

FUNDERBURK, Robert Steele.
History of Conservation Education in the United States (Nashville, TN, 1948).

GALE, R.P.
"Social Movements and the State: The Environmental Movement, Countermovement, and Government Agencies," *Sociological Perspectives* 29 (1986):202-40. (P.P.8044.QR)

GANOE, John T.
"The Origin of a National Reclamation Policy," *Mississippi Valley Historical Review* 18 (June 1931):34-52. (Ac.8408/2)

GOETZMANN, William H.
Looking Far North: The Harriman Expedition to Alaska, 1899 (Princeton, NJ: Princeton University Press, 1982), (X.809/56156)

GRAHAM, Frank, Jr.
The Adirondack Park: A Political History (New York: Knopf, 1978). (X.322/10217)
  Man's Dominion: The Story of Conservation in America (New York: Lippincott, 1971).
  Since Silent Spring (London: Hamilton, 1970). (X.329/4191)

HANDLEY, William.
Natural History in America: From Mark Catesby to Rachel Carson (New York: Quadrangle Books, 1977).

HAYS, Samuel P.
Beauty, Health, and Permanence: Environmental Politics in the United States, 1955-1985 [In collaboration with Barbara D. Hays] (Cambridge: Cambridge University Press, 1987). (YC.1987.b.4567)
  Conservation and the Gospel of Efficiency: The Progressive Conservation Movement, 1890-1920 (Cambridge, MA: Harvard University Press, 1959). (Ac.2692.bk)
  "From Conservation to Environmental Policies in the United States Since World War Two," *Environmental Review* 6 (1982):14-41. (Boston Spa)
  "Gifford Pinchot and The American Conservation Movement," in *Technology in America: A History of Individuals and Ideas,* Carroll W. Purcell, ed. (Cambridge, MA:

MIT Press, 1981).
"The Structure of Environmental Politics since World War II," *Journal of Social History* 14 (Summer 1981):719-738.

HIGHSMITH, Richard Morgan, J. Granville Jensen and Robert Dean Rudd.
Conservation in the United States Rand McNally Geography Series (Chicago, IL: Rand McNally and Co., 2nd ed., 1969). (X.322/1129)

HUGHES, J. Donald.
American Indian Ecology (El Paso, TX: Texas Western University Press, 1983).
Ecology in Ancient Civilizations (Albuquerque, NM: University of New Mexico Press, 1975). (X.809/45337)

HYNES, H. Patricia.
The Recurring Silent Spring (New York: Pergaman Press, 1989).

JACOBS, Wilbur R.
"The Great Despoliation: Environmental Themes in American Frontier History," *Pacific Historical Review* 47 (February 1978):1-26. (Ac. 8504.c)
"Frontiersmen, Fur Traders, and Other Varmints: An Ecological Appraisal of the Frontier in American History," *AHA Newsletter* 8 (November 1970):5-11. (P.P.8004.i)

JAMES, George Wharton.
Reclaiming the Arid West: The Story of the United States Reclamation Service (New York: Dodd and Mead, 1917). (X.809/56917)

KING, Judson.
The Conservation Fight: From Theodore Roosevelt to the Tennessee Valley Authority (Washington, D.C.: Public Affairs Press, 1959). (08248.s.8)

KINNEY, Jay P.
Indian Forest and Range: A History of the Administration and Conservation of the Redman's Heritage (Washington, D.C.: Forestry Enterprises, 1951).

KOPPES, Clayton R.
"Environmental Policy and American Liberalism: The Department of the Interior, 1933-1953," in *Environmental History: Critical Issues in Comparative Perspective*, Kendall E. Bailes, ed. (Lanham, MD: University Press of America, 1985). Originally published in *Environmental Review* 7 (1983):17-41. (Boston Spa)
"Public Waters, Private Land: Origins of the Acreage Limitation Controversy, 1938-1952," *Pacific Historical Review* 47 (1978):607-38. (Ac.8504.c)

LAMM, Richard D. and Michael McCarthy,
The Angry West: A Vulnerable Land and Its Furture (Boston, MA: 1982). [Lamm was Governor of the State of Colorado]

LASH, Jonathan.
A Season of Spoils: The Story of the Reagan Administration's Attack on the Environment (New York: Pantheon, 1984).

LENENSTEIN, Harvey A.
The Paradox of Plenty: A Social History of Eating in Modern America (New York: Oxford University Press, 1993). (YC.1993.b.3100)
Revolution at the Table: The Transformation of the American Diet (New York: Oxford University Press, 1988). (YC.1989.b.2681)

LEWIS, Martin W.
Green Delusions: An Environmentalist Critique of Radical Environmentalism (Durham, NC: Duke University Press, 1992). (YC.1993.b.2716)

LIEBER, Richard.
America's Natural Wealth: A Story of the Use and Abuse of Our Resources (New York: Harper, 1942). (10413.m.34)

MCCARTHY, George Michael.
Hour of Trial: The Conservation Conflict in Colorado and the West, 1891-1907 (Norman, OK: University of Oklahoma Press, 1977).

MCCONNELL, Grant.
"The Conservation Movement: Past and Present," *Western Political Quarterly* 7 (September 1954):463-78. (Ac.2690.rc)

MCCLOSKEY, Michael.
"Twenty Years of Change in the Environmental Movement: An Insiders View," in *American Environmentalism: The U.S. Environmental Movement, 1970-1990* Riley E. Dunlap and Angela G. Mertig, eds. (Philadelphia, PA: Taylor and Francis, 1992), pp. 77-88. (YC.1993.b.1277)
"Wilderness Movement at the Crossroads, 1945-1970," *Pacific Historical Review* 41 (August 1972):346-61. (Ac.8504.c)

MACKINLEY, Charles.
Uncle Sam in the Pacific Northwest: Federal Management of Natural Resources in the Columbia River Valley (Berkeley, CA: University of California Press, 1952). (Ac.2689.go.(16))

MARTIN, Calvin.
Keepers of the Game: Indian-Animal Relationships in the Fur Trade (Berkeley, CA: University of California Press, 1978). (X.809/43567) [See also Shepard Krech, III, ed., *Indians, Animals, and the Fur Trade: A Critique of Keepers of the Game* (Athens, GA: University of Georgia Press, 1981). (YA.1989.a.3532)]

MARTIN, Daniel.
Three Mile Island: Prologue or Epilogue? (Cambridge, MA: Ballinger, 1980). (X.622/13885)

MELOSI, Martin V.
Coping with Abundance: Energy and Environment in Industrial America (Philadelphia, PA: Temple University Press, 1985). (YA.1990.b.1930)
Garbage in the Cities: Refuse, Reform and the Environment, 1880-1980 (College Station, TX: Texas A & M University Press, 1981). (X.520/35324)
ed. *Pollution and Reform in American Cities, 1870-1930* (Austin, TX: University of Texas Press, 1980). (X.520/22060)
"Urban Pollution: Historical Perspective Needed," *Environmental Review* 3 (1979):37-45. (Boston Spa)

MERCHANT, Carolyn.
Ecological Revolutions: Nature, Garden and Science in New England (Chapel Hill, NC: University of North Carolina Press, 1989).
The Death of Nature: Women, Ecology and the Scientific Revolution (San Francisco, CA: Harper and Row, 1979). (X.529/44233)
Radical Ecology: Search for a Liveable World (London: Routledge, 1992). (YK.1993.b.3731)
ed. "Women and Environmental History," Special Issue of *Environmental Review* 8 (1984). (Boston Spa)

MILBRATH, Lester W.
Environmentalists: Vanguard for a New Society (Albany, NY: State University of New York Press, 1984). (YA.1989.b.7646)

MILBRATH, Lester W. and Frederick R. Inscho, eds.
The Politics of Environmental Policy (Beverly Hills, CA: Sage Publications, 1975). (X.319/16414 Woolwich)

MITCHELL, R. C.
"From Conservation to Environmental

Movement: The Development of the Modern Environmental Lobbies," in *Government and Environmental Politics*, Michael J. Lacey, ed. (Washington, D.C.: Wilson Center Press, 1989), pp. 81-113.

NASH, Roderick.
"The American Cult of the Primitive," *American Quarterly* 18 (1966):517-37. (Ac.2692.p/32)
American Environmentalism: Readings in Conservation History 3rd ed. (New York: McGraw-Hill, 1990). (YK.1990.a.7298)
ed. *Environment and Americans: The Problem of Priorities*, American Problem Studies Series (New York: Holt, Rinehart & Winston, 1972).

NICHOLAS, George P., ed.,
Holocene Human Ecology in Northeastern North America (New York: Plenum Press, 1988). (YK.1990.b.235)

NIXON, Edgar B., compiler and editor.
Franklin D. Roosevelt and Conservation, 1911-1945 2 vols. (Hyde Park, NY: Franklin D. Roosevelt Library, 1957). (7009.h.25)

NOGGLE, Burl.
Teapot Dome: Oil and Politics in the 1920s (Baton Rouge, LA: Louisiana State University Press, 1962). (X.708/2948)

OPIE, John, ed.
Americans and Environment: The Controversy over Ecology, Problems in American Civilization Series (Lexington, MA: D. C. Heath, 1971).

O'RIORDAN, T.
"The Third American Conservation Movement: New Implications for Public Policy," *Journal of American Studies* 5 (1971):155-71. (P.901/236)

PAEHLKE, Robert C.
Environmentalism and the Future of Progressive Politics (New Haven, CT: Yale University Press, 1989). (YC.1991.b.3978)

PAEHLKE, Robert C. and Douglas Torgerson, eds.
Managing Leviathan: Environmental Politics and the Administrative State (Lewiston, NY: Broadway Press, 1990). (YA,1993.a.12476)

PENICK, James.
Progressive Politics and Conservation: The Ballinger-Pinchot Affair (Chicago, IL: University of Chicago Press, 1968). (X.709/7272)

PERKINS, John H.
Insects, Experts, and the Insecticide Crisis: The Quest for New Pest Management Strategies (New York: Plenum, 1982).

PESKIN, Henry M, Paul R. Portney, and Allen V. Kneese, eds.,
Environmental Regulation and the U.S. Economy (Baltimore, MD: Published for Resources for the Future by Johns Hopkins University Press, 1981). (X.520/33916)

PETULLA, Joseph M.
American Environmental History: The Exploitation and Conservation of Natural Resources (San Francisco, CA: Boyd and Fraser Publishing Co., 1977).
American Environmentalism: Values, Tactics, Priorities (College Station, TX: Texas A&M University Press, 1980).

POMEROY, Earl.
In Search of the Golden West: The Tourist in Western America (New York: Knopf, 1957). (YH.1986.a.113)

PORTNEY, Kent E.
Controversies in Environmental Policy: Science

vs. Economics vs. Politics (Beverly Hills, CA: Sage Publications, 1992). (YC.1993.a.2128)

PORTNEY, Paul R., ed.
Current Issues in Natural Resource Policy (Washington, D.C.: Resources for the Future, 1982). (X.520/29722)
Natural Resources and the Environment: The Reagan Approach (Washington, D.C.: Urban Institute Press, 1984).

POWELL, John Wesley.
Report on the Lands of the Arid Region of the United States, with a more detailed account of the land of Utah. With Maps (Washington, D.C.: Department of the Interior, 1878). (A.S.206/6)

REICH, Charles.
The Greening of America (Harmondsworth: Penguin, 1971). (X.700/8161)

REIGER, John F.
American Sportsmen and the Origins of Conservation (1975; Norman, OK: University of Oklahoma Press, revised ed., 1986). (YC.1988.a.10454)

RICHARDSON, Elmo R.
Dams, Parks and Politics: Resource Development and Preservation in the Truman and Eisenhower Era (Lexington, KY: University of Kentucky Press, 1973).
The Politics of Conservation: Crusades and Controversies, 1897-1913 (Berkeley, CA: University of California Press, 1962).
"The Interior Secretary as Conservation Villain: The Notorious Case of Douglas 'Giveaway' McKay," *Pacific Historical Review* 41 (August 1972):333-45. (Ac.8504.c)

RIESCH-OWEN, A. L.
Conservation Under FDR (New York: Praeger, 1983). (X.800/36566)

ROSENKRANTZ, Barbara Gutmann, and William A. Koelsch, eds.
American Habitat: A Historical Perspecitive (New York: Free Press, 1973).

ROTHMAN, Hal.
Preserving Different Pasts: The American National Monuments (Urbana, IL: University of Illinois Press, 1989). (YA.1994.b.3411)

SALE, Kirkpatrick.
The Conquest of Paradise: Christopher Columbus and the Columbian Legacy (London: John Curtis book from Hodder and Stoughton, 1991). (YC.1991.b.1170)
Dwellers in the Land: The Bioregional Vision (San Francisco, CA: Sierra Club, 1981).

SALMOND, John A.
The Civilian Conservation Corps, 1933-1942: A New Deal Case Study. (Durham, NC: Duke University Press, 1967).

SAX, Joseph L.
Defending the Environment, A Strategy for Citizen Action (New York: Knopf, 1971). (X.329/6443)

SCARCE, Rik.
Eco-warriors: Understanding the Radical Environmental Movement (Chicago, IL: Noble Press, 1990). (YA.1992.a.13392)

SCHREPFER, Susan R.
The Fight to Save the Redwoods: A History of Environmental Reform, 1917-1978 (Madison, WI: University of Wisconsin Press, 1983). (X.322/13020)

SCHURR, Sam H. and Bruce C. Netschert.
Energy in the American Economy, 1850-1975: An Economic Study of its History and Prospects (Baltimore, MD: Johns Hopkins University Press for Resources for the Future, 1960). (8291.k.23)

SHELFORD, Victor E.
The Ecology of North America (1963; Urbana, IL: University of Illinois Press, 1978). (X.311/8905)

SMITH, Frank E.
The Politics of Conservation (New York: Pantheon Books, 1966). (X.709/7660)

STANNARD, David E.
American Holocaust: Columbus and the Conquest of the New World (New York: Oxford University Press, 1993). (YC.1993.b.4163)

STEGNER, Wallace E.
Beyond the Hundredth Meridian: John Wesley Powell and the Second Opening of the West (1954; Boston, MA: Houghton Mifflin, 1962). (X.809/1100)

STILGOE, John R.
Borderland: Origins of the American Suburb, 1820-1939 (New Haven, CT: Yale University Press, 1988). (YH.1989.b.930)
    Common Landscape of America, 1580 to 1845 (New Haven, CT: Yale University Press, 1982). (X.800/34098)

STRONG, Douglas H.
The Conservationists (New York: Addison-Wesley, 1971).
    Dreamers and Defenders: American Conservationists (Lincoln, NE: University of Nebraska Press, 1988).

SZASZ, Andrew.
EcoPopulism, Toxic Waste and the Movement for Environmental Justice (Minneapolis, MN: University of Minnesota Press, 1994).

TALBOT, Alan R.
Power Along the Hudson: The Storm King Case and the Birth of Environmentalism (New York: Dutton, 1972).

TAYLOR, Paul S.
"Reclamation: The Rise and Fall of an American Idea," *American West* 7 (1970):27-33,63. (P.P. 8003.zw)

TOBER, James A.
Who Owns the Wildlife?: The Political Economy of Conservation in Nineteenth-Century America (Westport, CT: Greenwood Press, 1981).

TURNER, Frederick W.
Beyond Geography: The Western Spirit Against the Wilderness (1980; New Brunswick, NJ: Rutgers University Press, 1983). (YA.1987.b.2544)
    Rediscovering America: John Muir in His Time and Ours (New York: Viking, 1978).

UDALL, Stewart L.
The Quiet Crisis (New York: Holt, Rinehart and Winston, 1963). [Udall was the Secretary of the Interior in the Kennedy Administration] (X.510/1145)
    The Quiet Crisis and the Next Generation (1963; Salt Lake City, UT: Peregrine Smith, 1988).

VAN HISE, Charles.
The Conservation of Natural Resources in the United States (New York: Macmillan, 1910). (08227.aa.22)

VECSEY, Christopher and Robert W. Venables, eds.
American Indian Environments: Ecological Issues in Native American History (Syracuse, NY: Syracuse University Press, 1980).

WENGERT, Norman.
Natural Resources and the Political Struggle (Garden City, NY: Doubleday, 1955).

WHISENHUNT, Donald W.
The Environment and the American Experience: A Historial Look at the Ecological Crisis (Port Washington, NY: Kennikat, 1974).

WHITAKER, John C.
Striking a Balance: Environment and Natural Resource Policy in the Nixon-Ford Years (Washington, D.C.: American Enterprise Institute for Public Policy Research and Stanford Hoover Institution on War, Revolution and Peace, 1976). (Ac.2692.nf/4(57))

WHITE, Edward G.
The Eastern Establishment and the Western Experience: The West of Frederic Remington, Theodore Roosevelt and Owen Wister (New Haven, CT: Yale University Press, 1968). (W.P.4495/14)

WHITE, Richard.
"Animals and Enterprise," in *The Oxford History of the American West*, Clyde A. Milner II, Carol A. O'Connor and Martha A. Sandweiss, eds. (New York: Oxford University Press, 1994), pp. 237-73.

It's Your Own Misfortune and None of My Own: A History of the American West (Norman OK: University of Oklahoma Press, 1991). (YC.1993.b.406)

Land Use, Environment, and Social Change: The Shaping of Island County, Washington (Seattle, WA: University of Washington Press, 1980), (X.520/23086)

"Native Americans and the Environment," in *Scholars and the Indian Experience: Critical Reviews of Recent Writing in the Social Sciences* ed. W.R. Swagerty (Bloomington, ID: Indian University Press, 1984), pp. 179-204. (YA.1990.b.1878)

The Roots of Dependency: Subsistence, Environment and Social Change Among the Choctaws, Pawnees and Navajos (Lincoln, NE: University of Nebraska Press, 1983). (X.520/32274)

WHORTON, James C.
Before 'Silent Spring': Pesticides and Public Health in Pre-DDT America (Princeton, NJ: Princeton University Press, 1974).

WORSTER, Donald. ed.
American Environmentalism: The Formative Period, 1860-1915 (New York: Wiley, 1973).

Dust Bowl: The Southern Plains in the 1930s (New York: Oxford University Press, 1979). (X.800/15315)

ed. *The Ends of the Earth: Perspectives on Modern Environmental History* (Cambridge, UK: Cambridge University Press, 1988). (YC.1989.b.3346)

Nature's Economy: A History of Ecological Ideas (Cambridge: Cambridge University Press, 1985). (YH.1989.b.304)

Rivers of Empire: Water, Aridity and the Growth of the American West (1985; Oxford, UK: Oxford University Press, 1992). (YK.1992.b.6568)

Under Western Skies: Nature and History in the American West (New York: Oxford University Press, 1992). (YC.1992.b.3334)

The Wealth of Nature: Environmental History and the Ecological Imagination (New York: Oxford University Press, 1993). (YC.1993.b.5892)

WILDLIFE AND ANIMAL RIGHTS

ALLEN, Thomas B.
Guardian of The Wild: The Story of The National Wildlife Federation, 1936-1986 (Indianapolis, IN: Indiana University Press, 1987).

BEAN, Michael J.
The Evolution of National Wildlife Law (New York: Praeger, Rev. ed., 1983).
(X.322/13374)

CART, Theodore Whaley.
"The Struggle for Wildlife Protection in the United States, 1870-1900: Attitudes and Events Leading to the Lacey Act," (Ph.D. dissertation, University of North Carolina at Chapel Hill, 1971).

DUNLOP, Thomas P.
Saving America's Wildlife: Ecology and the American Mind, 1850-1990 (Princeton, NJ: Princeton University Press 1988).
(YC.1991.b.6537)

FLADER, Susan L.
Thinking Like a Mountain: Aldo Leopold and the Evolution of an Ecological Attitude toward Deer, Wolves, and Forests (1974; Lincoln, NE: University of Nebraska Press, 1978).
(X.319/18692)

GIBBONS, Felton and Deborah Strom.
Neighbors to the Birds: A History of Birdwatching in America (New York: Norton, 1988).

GRAHAM, Edward H.
The Land and Wildlife (New York: Oxford University Press, 1947). (7080.aaa.48)
and William R. Van Dersal. *Wildlife for America: The Story of Wildlife Conservation* (New York: Oxford University Press, 1949).
(07209.cc.33)

KELLERT, Stephen R.
Policy Implications of a National Study of American Attitudes and Behavioral Relations to Animals (Washington, D.C.: Government Printing Office for the Department of Interior Fish and Wildlife Service, 1978).
(A.S. 193/185)

LEAVITT, Emily S.
Animals and Their Legal Rights: A Survey of American Laws from 1641 to 1978. (Wahington, D.C., 1978).

LEEDS, Anthony, and Andrew P. Vayda, eds.
Man, Culture, and Animals: The Role of Animals in Human Ecological Adjustments (Washington, DC: American Association for the Advancement of Science, 1965).

LUND, Thomas A.
American Wildlife Law (Berkeley, CA: University of California Press, 1980).
(X.200/34064)

MCEVOY, Arthur F.
The Fisherman's Problem: Ecology and Law in the California Fisheries, 1850-1980 (Cambridge: Cambridge University Press, 1986). (YK.1987.b.703)

MCHUGH, Tom.
The Time of the Buffalo (Lincoln, NE: University of Nebraska Press, 1972).

MAGEL, Charles R.
A Bibliography of Animals Rights and Related Matters (Lanham, MD: University Press of America, 1981). [3,200 items in 622 pp.]
Keyguide to Information Sources in Animal Rights (London: Mansell, 1989).
(YK.1989.b.1625)

MAGEL, Charles R. and Tom Regan.
"A Selected Bibliography on Animal Rights and Human Obligations," *Inquiry* 22 (Summer 1979):243-47.

MATTHIESSEN, Peter.
Wildlife in America (New York: Penguin, 1977). (X.319/18812)

MIGHETTO, Lisa.
Wild Animals and American Environmental Ethics (Tucson, AZ: University of Arizona Press, 1991).
ed. *Muir Among the Animals: The Wildlife Writings of John Muir* (San Francisco, CA: Sierra Club Books, 1986).

ORR, Oliver H.
Saving American Birds: T. Gilbert Pearson and the Founding of the Audubon Movement (Gainesville, FL: University of Florida Press, 1992). (YC.1993.a.1794)

PALMER, William D.
"Endangered Species Protection: A History of Congressional Action," *Environmental Affairs* 4 (Spring 1975):255-293.

PARKES, Patrick B. and Jacques V. Sichel.
The Humane Society of the United States, 1954-1979: Twenty-Five Years of Growth and Achievement (Washington, D.C., 1978).

REED, Nathaniel P. and Dennis Drabelle.
The United States Fish and Wildlife Service (Boulder, CO: Westview, 1984). (X.322/15542)

REGAN, Tom, ed.,
All That Dwells Therein: Animal Rights and Environmental Ethics (Berkeley, CA: University of California Press, 1982). (X.529/51733)
The Case for Animal Rights (London: Routledge and Kegan Paul, 1983). (X.322/13945)
Earthbound: New Introductory Essays in Environmental Ethics (New York: Random House, 1984).

ROE, Frank Roe.
The North American Buffalo: A Critical Study of the Species in Its Wild State (Toronto: University of Toronto Press, 1951). (07209.dd.3)

RYDER, Richard D., ed.
Welfare and the Environment (London: Duckworth, 1992). (YK.1993.a.2457)

SCHORGER, Arlie William.
The Passenger Pigeon: Its Natural History and Extinction (Norman, OK: University of Oklahoma Press, 1955)

SHERWOOD, Morgan B.
Big Game in Alaska: A History of Wildlife and People (New Haven, CT: Yale University Press, 1981). (AC.2692.ma/32)

SHOEMAKER, Carl D.
The Stories Behind the Organization of the National Wildlife Federation (Washington D.C.: National Wildlife Federation, 1960).

TREFETHEN, James B.
An American Crusade for Wildlife: Highlights in Conservation Progress (New York: Winchester, 1975).

WATKINS, T.H.
Vanishing Arctic: Alaska's National Wildlife Refuge (New York: Aperture/Wilderness Society, 1988).
(L.B. 31.b.3557)

WISHART, David.
The Fur Trade of the American West, 1807-1840: A Geographical Synthesis (Lincoln, NE: University of Nebraska Press, rev. ed., 1993). (YK.1993.a.13451)

YAFFEE, Steven L.
Prohibitive Policy: Implementing the Federal Endangered Species Act (Cambridge, MA: MIT Press, 1982). (X.520/28535)

WATER, SOIL AND MINERALS
[See also Regional Studies]

BENNETT, Hugh Hammond.
Elements of Soil Conservation (New York: McGraw-Hill, 1947). (2249.c.20)

BLAKE, Nelson M.
Land into Water, Water into Land: A History of Water Management in Florida (Tallahassee, FL: University Presses of Florida, 1980).

BOWEN, Charles.
Killing the Hidden Waters (Austin, TX: University of Texas Press, 1977).

FITE, Gilbert.
American Farmers: The New Minority (Bloomington, IN: Indiana University Press, 1981). (X.520/29705)
　　The Farmer's Frontier, 1865-1900, Histories of the Frontier Series (New York: Holt, Rinehart and Winston, 1966). (X.311/1301)

FRADKIN, Philip L.
A River No More: The Colorado River and the West (Tucson, AZ: University of Arizona Press, 1984). (YA.1989.a.13301)

FRANK, Arthur D.
The Development of the Federal Program of Flood Control on the Mississippi River (New York: Columbia University Press, 1930).

FRANK, Bernard and Anthony Netboy.
Water, Land and People (New York: Knopf, 1950).

GRAY, Lewis Cecil.
History of Agriculture in the Southern United States to 1860 (Washington, D.C.: Carnegie Institution, Publication no. 430, 1935). (Ac.1866)

HARDIN, Charles Meyer.
The Politics of Agriculture: Soil Conservation and the Struggle for Power in Rural America (Glencoe, IL: Free Press, 1952). (7082.e.3)

HARGREAVES, Mary W. M.
Dry Farming in the Northern Great Plains, 1900-1925 (Cambridge, MA: Harvard University Press, 1957). (Ac.2692/11)

HELD, R. Burnell, and Marion Clawson.
Soil Conservation in Perspective (Baltimore, MD: Johns Hopkins University Press, 1965). (X.311/1699)

HUFFMAN, Roy E.
Irrigation Development and Public Water Policy (New York: Ronald, 1953).

HUNDLEY, Norris.
Dividing the Waters: A Century of Controversy Between the United States and Mexico (Berkeley, CA: University of California Press, 1966). (X.709/3629)
　　The Great Thirst, California and Water, 1770s-1990s (Berkeley, CA: University of California Press, 1992). (YK.1993.b.4025)
　　Water and the West: The Colorado River Compact and the Politics of Water in the American West (Berkeley, CA: University of California Press, 1975). (X.809/40937)

ISE, John.
The United States Oil Policy (New Haven, CT: Yale University Press, 1926). (8228.pp.15)

JACKS, Graham V. and Robert O. Whyte.
The Rape of the Earth: A World Survey of Soil Erosion (London: Faber and Faber, 1939). (07076.c.15)

KAHRL, William L., ed.
The California Water Atlas (Sacramento, CA: 1979).
Water and Power: The Conflict over Los Angeles' Water Supply in the Owens Valley (Berkeley, CA: University of California Press, 1982). (X.622/13582)

KERWIN, Jerome.
Federal Water-Power Legislation (New York: Columbia University Studies in History, Economics, and Public Law, No. 274, 1926). (Ac.2688/2)

MAASS, Arthur.
Muddy Waters: The Army Engineers and the Nation's Rivers (Cambridge, MA: Harvard University Press, 1951). (Ac.2692.am/22)

MCDONALD, Angus.
Early American Soil Conservationists (Washington, D.C.: U.S. Government Printing Office, 1941).

MALIN, James C.
History and Ecology: Studies of The Grasslands. Introduction by Robert P. Swierenga (Lincoln, NE: University of Nebraska Press, 1984). (YA.1989.b.4480)
The Grassland of North America: Prolegomena to Its History With Addenda and Postscripts (Lawrence, KA: James C. Malin, 1961). (X.319/1451)

MORGAN, Murray C.
The Columbia: Powerhouse of the West (Seattle, WA: Superior Publishing Co., 1949). (10414.e.35)

MORGAN, Robert J.
Governing Soil Conservation: Thirty Years of New Decentralization (Baltimore, MD: Johns Hopkins University Press, 1965).

NETBOY, Anthony.
The Columbia River Salmon and Steelhead Trout: Their Fight for Survival (Seattle, WA: University of Washington Press, 1980). (X.809/47837)

PALMER, Tim.
Endangered Rivers and the Conservation Movement (Berkeley, CA: University of California Press, 1986).
Stanislaus: The Struggle for a River (Berkeley, CA: University of California Press, 1982). (X.322/12096)

PISANI, Donald J.
From the Family Farm to Agribusiness: The Irrigation Crusade in California and the West, 1850-1931 (Berkeley, CA: University of California Press, 1984). (X.322/14987)
"Forests and Conservation, 1865-1890," *Journal of American History* 72 (September 1985):340-59. (Ac. 8408/2)
"The Irrigation District and the Federal Relationship, Neglected Aspects of Water History in the Twentieth-Century West," in *The Twentieth-Century West, Historical Interpretations*, Gerald D. Nash and Richard W. Etulain, eds., (Albuquerque, NM: University of New Mexico Press, 1989), pp. 257-92. (YA.1990.b.150)
To Reclaim a Divided West: Water Law and Public Policy, 1848-1902 (Albuquerque, NM: University of New Mexico Press, 1992). (YA.1993.b.5781)

PURSELL, Carroll, ed.
From Conservation to Ecology (New York: Cromwell, 1979).

RAUP, Hugh M.
"Some Problems in Ecological Theory and Their Relation to Conservation," *Journal of Ecology* 52 (Supplement, 1964), pp. 19-28.

REISNER, Marc.
Cadillac Desert: The American West and Its Disappearing Water (1986; London: Secker and Warburg, 1990). (YC.1990.b.1664)

RICKARD, Thomas A.
History of American Mining (New York: McGraw-Hill, 1932). (Ac.3238.g/2(3))

RUDD, Robert L.
Pesticides and the Living Landscape (Madison, WI: University of Wisconsin Press, 1964).

SMITH, Duane A.
Mining America: The Industry and the Environment (Lawrence, KS: University of Kansas Press, 1987).

SMYTHE, William E.
The Conquest of Arid America (New York: Harper and Bros., 1900). (07078.de.76). [1969 edition: Lawrence B. Lee, ed., Seattle, WA: University of Washington Press, (X.809/7687)]

SPENCE, Clark C.
"The Golden Age of Dredging: The Development of an Industry and Its Environmental Impact," *Western Historical Quarterly* 11 (October 1980):401-14. (P.701/404)

STEVENS, Joseph E.
Hoover Dam: An American Adventure (Norman, OK: University of Oklahoma Press, 1988). (YC.1989.b.1739)

SUNDBORG, George.
Hail Columbia: The Thirty-Year Struggle for Grand Coulee Dam (New York: Macmillan, 1954).

TOBEY, Ronald C.
Saving the Prairies: The Life Cycle of the Founding School of American Plant Ecology, 1895-1955 (Berkeley, CA: University of California Press, 1981). (X.322/11022)

TYLER, Daniel.
The Last Water Hole in the West: The Colorado-Big Thompson Project and the Northern Water Conservancy District (Niwot, CO: University Press of Colorado, 1992).

VIETOR, Richard H.K.
Environmental Politics and the Coal Coalition (College Station, TX: Texas A & M University Press, 1980).
"The Evolution of Public Environmental Policy: The Case of 'Non-Significant Deterioration'," *Environmental Review* 3 (Winter 1979):3-19. (Boston Spa)

WARNE, William.
The Bureau of Reclamation (New York: Praeger, 1973).

WHEELER, Donald.
TVA and the Tellico Dam, 1936-1979 (Knoxville, TN: University of Tennessee Press, 1986).

ZIMMERMANN, Erich W.
Conservation in the Production of Petroleum: A Study of Industrial Control (New Haven, CT: Yale University Press, 1957).

PARKS AND PUBLIC LAND
[See also Forests]

ALBRIGHT, Horace and Robert Cahn.
The Birth of the National Park Service: The Founding Years (Salt Lake City, UT: Howe, 1985).

ALEXANDER, Thomas G.
"Senator Reed Smoot and Western Land Policy, 1905-1920," *Arizona and the West* 13 (Autumn 1971):245-64. (P.701/1302)

BARNES, Wil C.
Western Grazing Grounds and Forest Ranges (New York: Arno Press, 1979). (YA.1990.a.22532)

BARTLETT, Richard A.
Yellowstone: A Wilderness Besieged (Tucson, AZ: University of Arizona Press, 1985).
Nature's Yellowstone (Albuquerque, NM: University of New Mexico Press, 1974). (X.800/13423)

BUCKHOLTZ, C.W.
Rocky Mountain National Park: A History (Boulder, CO: Colorado Associated University Press, 1983). (YA.1989.b.4320)

CAMERON, Jenks.
The National Park Service: Its History, Politics and Organization (New York: Appleton, 1922). (Ac.1869/3)

CARSTENSEN, Vernon, ed.
The Public Lands: Studies in the History of the Public Domain (Madison, WI: University of Wisconsin Press, 1963). (X.200/1728)

CAWLEY, R. McGreggor. *Federal Land, Western Anger: The Sagebrush Rebellion and Environmental Politics* (Lawrence, KA: University Press of Kansas, 1993). (YC.1994.b.2036)

CHASE, Alston.
Playing God in Yellowstone: The Destruction of America's First National Park (Boston, MA: Atlantic Monthly Press, 1986).

CLAWSON, Marion.
The Bureau of Land Management (New York: Praeger, 1971). (X.329/6397)
The Land System of the United States: An Introduction to the History and Practice of Land Use and Land Tenure (Lincoln, NE: University of Nebraska Press, 1968). (X.329/4388)
Man and Land in the United States (Lincoln, NE: University of Nebraska Press, 1964). (X.809/3495)

CLAWSON, Marion, and R. Burnell Held.
The Federal Lands: Their Use and Management (Baltimore, MD: Johns Hopkins University Press, 1957).
and Charles H. Stoddard. *Land for the Future* (Baltimore, MD: Johns Hopkins University Press for Resources for the Future, 1960). (7085.de.14)

COX, Thomas R.
"From Hot Springs to Gateway: The Evolving Concept of Public Parks, 1832-1976," *Environmental Review* 5 (1981):14-26. (Boston Spa)
The Park Builders: A History of State Parks in the Pacific Northwest (Seattle, WA: University of Washington Press, 1988). (YC.1990.a.1136)

CULHANE, Paul J.
Public Lands Politics: Interest Group Influence on the Forest Service and the Bureau of Land Management (Baltimore, MD: Johns Hopkins University Press, 1981).

EARLY, Katherine E.
"For the Benefit and Enjoyment of the People": Cultural Attitudes and the Establishment of

Yellowstone National Park (Washington, DC: Georgetown University Press, 1984). (YA.1987.a.15516)

EVERHART, William C.
The National Park Service (New York: Praeger, 1972). (X.622/16335)

FORESTA, Ronald A.
America's National Parks and Their Keepers (Washington, D.C.: Resources for the Future, 1984). (X.322/14804)

FOSS, Phillip O.
Politics and Grass: The Administration of Grazing on the Public Domain (Seattle, WA: University of Washington Press, 1960).

FRANCIS, John G. and Richard Ganzell, eds.
Western Public Lands: The Management of Natural Resources in the Time of Declining Federalism (Totowa, NY: Rowman and Allanheld, 1984). (YA.1990.a.22532)

FREEMUTH, John C.
Islands Under Siege: National Parks and the Politics of External Threats (Lawrence, KS: University Press of Kansas, 1991). (YA.1993.b.7709)

GATES, Paul W.
History of Public Land Law Development (1968; Washington, D.C.: Zenger Publishing Co., 1978). (YA.1986.b.1157).

HAINES, Aubrey.
The Yellowstone Story: A History of Our First National Park 2 volumes (Yellowstone, WY: Yellowstone Library and Museum Association, 1977). (X.322/15029)

HAMPTON, H. Duane.
"Opposition to National Parks," *Journal of Forest History* 25 (January 1981):36-45.
How The U.S. Cavalry Saved Our National Parks (Bloomington, IN: Indiana University Press, 1971).

HANDEE, John, George H. Stonkay, and Robert C. Lucus.
Wilderness Management (Washington D.C.: U.S. Government Printing Office, 1978).

HIBBARD, Benjamin H.
A History of Public Land Policies (New York: Macmillan, 1924). (X.525/907(3))

ISE, John.
Our National Parks: A Critical History (Baltimore, MD: Johns Hopkins University Press, 1961). (010160.k.32)

JOHNSON, Hildegard Binder.
Order Upon the Land: The U.S. Rectangular Land Survey and the Upper Mississippi Country (New York: Oxford University Press, 1976). (X.800/25610)

JONES, Holway R.
John Muir and the Sierra Club: The Battle for Yosemite (San Francisco, CA: Sierra Club, 1965). (X.802/1020)

NASH, Roderick.
"The American Invention of National Parks," *American Quarterly* 22 (Fall 1970):726-35. (Ac. 2692.p/32)
"Tourism, Parks and the Wilderness Idea in the History of Alaska," *Alaska in Perspective* 4 (1981):1-27.

PEFFER, E. Louise.
The Closing of the Public Domain: Disposal and Reservation Policies, 1990-1950 (Stanford, CA: Stanford University Press, 1951). (Ac.2692.nh)

PYNE, S.J.
Fire in America; A Cultural History of Wildland and Rural Fire (Princeton, NJ: Princeton University Press, 1982). (X.800/34099)

Introduction to Wildland Fire: Fire Management in the United States (New York: Wiley, 1984). (X.622/20935)

RAKESTRAW, Lawrence.
"The West, States' Rights, and Conservation: A Study of Six Public Land Conferences," *Pacific Northwest Quarterly* 48 (July 1957):89-99. (P.P. 8004.jv)

RIGHTER, Robert.
Crucible for Conservation: The Creation of Teton National Park (Boulder, Co: Colorado Associated University Press, 1982).

ROBBINS, Roy M.
Our Landed Heritage: The Public Domain, 1776-1936 (Princeton NJ: Princeton University Press, 1942). (8287.f.47)

ROWLEY, William D.
U.S. Forest Service Grazing and Rangelands, A History (College Station, TX: Texas A & M University Press, 1985). (YA.1990.b.1348)

RUNTE, Alfred.
National Parks: The American Experience (Lincoln, NE: University of Nebraska Press, 1979). (X.429/10612) [(2nd ed., Linclon, NE: University of Nebraska Press, 1987)]
  Yosemite: The Embattled Wilderness (Lincoln, NE: University of Nebraska Press, 1990). (YK.1994.b.1554)
  ed. *Yosemite and Sequoia: A Century of California National Parks*, [co-edited with Richard J. Orsi and Marlene Smith-Barangini] (Berkeley, CA: University of California Press, 1993). (YC.1993.b.7690)

RUSSELL, Carl Parcher.
One Hundred Years in Yosemite (Stanford, CA: Stanford University Press, 1931). (20016.cc.24)

SAX, Joseph L.
Mountains Without Handrails: Reflections on the National Parks (Ann Arbor, MI: University of Michigan Press, 1980).

SHORT, Calvin Brant.
Ronald Regan and the Public Lands: America's Conservation Debate, 1979-1984 (College Station, TX: Texas A & M University Press, 1989). (YA.1990.a.7606)

STRONG, Douglas H.
Trees or Timber?: The Story of Sequoia and Kings Canyon National Parks (Three Rivers, CA: Sequoia Natural History Association, 1968).

THOMPSON, Gregory.
Parks in the West and American Culture (San Valley, ID: Institute of the American West, 1984). Includes a bibliography of histories of national parks.

TILDEN, Freeman.
The State Parks, Their Meaning in American Life (New York: Knopf, 1962). (X.311/797)

TWIGHT, Ben W.
Organizational Values and Political Power: The Forest Service Versus the Olympic National Park (University Park, PA: Pennsylvania Press, 1983). (X.529/67085)

UDALL, Stewart L.
The National Parks of America (Waukesha, MI: Country Beautiful Corp., 1972). (X.325/135)

VOIGHT, William, Jr.
Public Grazing Lands: Use and Misuse by Industry and Government (New Brunswick, NJ: Rutgers University Press, 1976).

WATKINS, T.W. and Charles S. Watkins, Jr.,
The Lands No One Knows: America and the Public Domain (San Francisco, CA: Sierra Club Books, 1975). (X.802/11110)
The Western Range, United States Senate,

Document 199, 74th Congress, 2d Session, Serial Number 10005 (Washington, DC: Government Printing Office, 1936). [A 620-page report of a federal survey of about 40% of the total land area of the continental United States, providing a detailed ecological analysis of the impact of western livestock grazing]

WYANT, William K.
Westward in Eden: The Public Lands and the Conservation Movement (Berkeley, CA: University of California Press, 1982). (X.800/33621)

ZASLOWSKY, Dyan.
These American Lands: Parks, Wilderness and the Public Lands (New York: Holt, 1986).

FORESTS [SEE ALSO PARKS AND PUBLIC LAND]

CAMERON, Jenks.
The Development of Governmental Forest Control in the United States (Baltimore, MD: Johns Hopkins University Press, 1928).

CARHART, Arthur H.
The National Forest (New York: Knopf, 1959).

CLARY, David A.
Timber and the Forest Service (Lawrence, KA: University of Kansas Press, 1986). (YK.1988.b.6205)

CLEPPER, Henry.
Crusade for Conservation: The Centennial History of the American Forestry Association (Washington D.C.: American Forestry Association, 1975).

CLEPPER, Henry and Arthur B. Meyer, eds.
American Forestry: Six Decades of Growth (Washington, D.C.: Society of American Forestry, 1960). (X.311/483)

COX, Thomas R., Robert S. Maxwell, Phillip Drennon Thomas, and Joseph J. Malone,
This Well-Wooded Land: Americans and their Forests from Colonial Times to the Present (Lincoln NE: University of Nebraska Press, 1985).

DANA, Samuel Trask.
Forest and Range Policy: Its Development in the United States (New York: McGraw-Hill, 1956). (W.P. 10970/37) [(2nd ed., with Sally K. Fairfax, New York, 1980)]

FROME, Michael.
Battle for the Wilderness (Epping: Bowker, 1984). (X.329/19037)
Conscience of a Conservationist: Selected

Essays (Knoxville, TN: University of Tennessee Press, 1989). (YA.1990.a.19795)
The Forest Service (1971; Boulder, CO: Westview, 2nd, revised ed., 1984). (X.322/14513)
ed. *Issues in Wilderness Management* (Boulder, CO: Westview, 1985). (X.329/20250)
The National Forests of America (Waukesha, WI: Country Beautiful Corp., 1968). (X.322/2410)
Strangers in High Places: The Story of the Great Smokey Mountains (Garden City, NY: Doubleday, 1966).
Whose Woods These Are: The Story of the National Forests (1963; Boulder, CO: Westfield, 1984). (X.329/19572)

GILLIGAN, James,
"The Development of Policy and Administration of Forest Service Primitive and Wilderness Areas in the Western United States," (Ph.D. dissertation, University of Michigan, 1953)

GREELEY, William B.
Forests and Men (Garden City, NY: Doubleday, 1951).
Forest Policy, American Forestry Series (New York: McGraw-Hill Book Co., 1953). (W.P. 10970/33)

HOUGH, Franklin B.
Report on Forestry, 4 volumes (Washington, D.C.: Forest Service, Department of Agriculture, 1878-1884) [vols. 1-3 by Hough; vol. 4 by W. H. Egelston] (A.S.822/2)

ISE, John.
The United States Forest Policy (New Haven, CT: Yale University Press, 1920). (7073.bb.44)

LOCKMAN, Ronald F.
Guarding the Forests of Southern California: Evolving Attitudes Toward Conservation of Watershed, Woodlands and Wilderness (Glendale, CA: Arthur H. Clark, 1981).

LILLARD, Richard G.
The Great Forest [The influence of forests on the history of the United States] (New York: Alfred A. Knopf, 1947). (7080.ee.9.)

LYON, Thomas J.
"American Nature Writing: A Selective Booklist on Nature and Man-in-Nature," *Antaeus* no. 57 (Autumn 1986):302-17. (Boston Spa)
ed. *This Incomparable Lande: A Book of American Nature Writing* (Boston MA: Houghton Mifflin, 1989).

OLSON, Sherry H.
The Depletion Myth: A History of Railroad Use of Timber (Cambridge, MA: Harvard University Press, 1971). (X.329.5342)

PLATT, Rutherford.
The Great American Forest (Englewood Cliffs, NJ: Prentice-Hall, 1965). (X.319/4938)

RAKESTRAW, Lawrence.
"A History of Forest Conservation in the Pacific Northwest, 1891-1915," (Ph.D. dissertation, University of Washington, 1955).
A History of the United States Forest Service in Alaska (Anchorage, AK: Alaska Historical Commission . . . , 1981).

ROBBINS, William G.
American Forestry: A History of National, State and Private Cooperation (Lincoln, NE: University of Nebraska Press, 1985).
Lumberjacks and Legislators: Political Economy of the U.S. Lumber Industry, 1890-1941 (College Station, TX: Texas A & M University Press, 1982). (X.800/40582)

ROBINSON, Glen O.
The Forest Service: A Study in Public Land

Management (Baltimore, MD: Johns Hopkins University Press, 1975).

ROTH, Dennis M.
The Wilderness Movement and the National Forests, 1964-1980 (Washington, D.C.: U. S. Government Printing Office, 1984).

SARGENT, Charles Sprague.
Report on the Forests of North America (Surveyed forestlands of United States, volume 9 of the 1880 U.S. Census) (Washington, D.C.: Tenth Census of United States, Census Office, 1883). (A.S. 70[10]8 & Maps.98.c.20)

SCHIFF, Ashley L.
Fire and Water: Scientific Heresy in the Forest Service (Cambridge, MA: Harvard University Press, 1962). (7083.n.28)

SHARP, Paul F.
"The Tree Farm Movement: Its Origin and Development," *Agricultural History* 23 (January 1949):41-45. (Ac.3510.a)

SHEPHERD, Jack.
The Forest Killers: The Destruction of the American Wilderness (New York: Weybright and Talley, 1975).

SMITH, Herbert A.
"The Early Forestry Movement in the United States," *Agriculture History* 12 (1938):326-46. (Ac.3510.a)

STEEN, Harold K.
The U.S. Forest Service: A History (Seattle, WA: University of Washington Press, 1976). (X.520/10667)
    ed. *The Origins of the National Forest: A Centennial Symposium* (Durham, NC: Forest History Society, 1992). (YA.1993.a.12297)

STEEN, Harold K. and Richard P. Tucker.
Changing Tropical Forests (Durham, NC: Forest History Society, 1992). (YK.1993.a.3518)

SWAIN, Donald C.
Forest Conservation Policy, 1921-33 (Berkeley, CA: University of California Press, 1963).

WILLIAMS, Michael.
Americans and Their Forests: A Historical Geography (Cambridge: Cambridge University Press, 1989). (YC.1990.b.3746)

REGIONAL STUDIES

ASHFORTH, William.
The Late, Great Lakes: An Environmental History (New York: Knopf, 1986). (YA.1993.a.19236)

BAMFORTH, Douglas B.
Ecology and Human Organization on the Great Plains (New York: Plenum, 1988). (YC.1989.b.6859)

BERGER, Jonathan, and John W. Sinton.
Water, Earth and Fire: Land Use and Environmental Planning in the New Jersey Pine Barrens (Baltimore, MD: Johns Hopkins University Press, 1985). (YK.1988.b.4561)

BLOUET, Brian W. and Frederick C. Luebke, eds.
The Great Plains: Environment and Culture (Lincoln, NE: University of Nebraska Press, 1979).

CLAPP, Gordon R.
The TVA: An Approach to the Development of a Region (Chicago, IL: University of Chicago Press, 1955). (Ac.2691.dw(28))

CLARK, Ira G.
Water in New Mexico: A History of Its Management and Use (Albuquerque, NM: University of New Mexico Press, 1987). (YA.1989.b.6271)

CLARK, Thomas D.
The Greening of the South: The Recovery of Land and Forest (Lexington, KY: University Press of Kentucky, 1984). (X.322/15035)

COATES, Peter.
The Trans-Alaska Pipeline Controversy: Technology, Conservation and the Frontier (Bethlehem, PA: Lehigh University Press, 1991). (YA.1993.b.9006)

COWDREY, Albert E.
This Land, This South: An Environmental History (Lexington, KY: University of Kentucky Press, 1983). (X.800/38875)

DASMANN, Raymond F.
California's Changing Environment (San Francisco, CA: Boyd and Fraser, 1981).
The Destruction of California (New York: Collier Books, Macmillan, 1965).

DEBUYS, William.
Enchantment and Exploitation: The Life and Hard Times of a New Mexico Mountain Range (Albuquerque, NM: University of New Mexico Press, 1985).

DOBYNS, Henry F.
From Fire to Flood: Historic Human Destruction of Sonoran Desert Riverine Oases (Socorro, NM: Ballena Press, 1981). (X.805/2201)

DOUGHTY, Robin W.
Wildlife and Man in Texas: Environmental Change and Conservation (College Station, TX: Texas A & M University Press, 1983).

DROZE, William A.
Trees, Prairies, and People: A History of Tree Planting in the Plains States (Denton, TX: Texas Women's University Press, 1977).

ENGLE, J. Ronald.
Sacred Sands: The Struggle for Community in the Indiana Dunes (Middletown, CT: Wesleyan University Press, 1983). (YA.1988.b.6204)

EVERS, Alf.
The Catskills: From Wilderness to Woodstock (Garden City, NY: Doubleday, 1972).

FARQUHAR, Francis P.
History of the Sierra Nevada (Berkeley, CA: University of California Press and the Sierra Club, 1965). (X.802/617)
Yosemite, the Big Trees and the High Sierra: A Selective Bibliography (Berkeley, CA:

University of California Press, 1948).
(11926.d.14)

FLADER, Susan L., ed.
The Great Lakes Forest: An Environmental and Social History (Minneapolis, MN: University of Minnesota Press, 1983).

FOX, Tom, Ian Koeppel, and Susan Kellam.
Struggle for Space: The Greening of New York City, 1970-1984 (New York: Neighborhood Open Space Coalition, 1985).

FRANKLIN, Kay, and Norman Schaeffer.
Duel for the Dunes: Land Use Conflict on the Shores of Lake Michigan (Urbana, IL: University of Illinois Press, 1983).

GREEN, Donald E.
Land of the Underground Rain: Irrigation on the Texas High Plains, 1910-1970 (Austin, TX: University of Texas Press, 1973).
(X.320/4125)

HART, Henry C.
The Dark Missouri (Madison, WI: University of Wisconsin Press, 1957). (010410.l.38)

HOFFMAN, Abraham.
Vision or Villainy? Origins of the Owens Valley-Los Angeles Water Controversy (College Station, TX: Texas A & M University Press, 1981). (X.800/41014)

HOLLON, W. Eugene.
The Great American Desert: Then and Now (New York: Oxford University Press, 1946).
(X.809/2448)
The Southwest: Old and New (New York: Alfred A. Knopf, 1961). (X.800/423)

HUBBARD, Preston J.
Origins of the TVA: The Muscle Shoals Controversy, 1920-1932 (Nashville, TN: Vanderbilt University Press, 1961).
(X.519/852)

HURT, R. Douglas.
The Dust Bowl: An Agricultural and Social History (Chicago, IL: Nelson-Hall, 1981).

INGERSOLL, William T.
"Land of Change: Four Parks in Alaska," *Journal of the West* 7 (April 1968):173-92.
(P.701/1257)

KEITER, Robert B. and Mark S. Boyce, eds.
The Greater Yellowstone Ecosystem; Redefining America's Wilderness Heritage (New Haven, CT: Yale University Press, 1991).
(YC.1992.b.4109)

KELLER, Jane Eblen.
Adirondack Wilderness: A Story of Man and Nature (Syracuse, NY: Syracuse University Press, 1980) (X.800/38877)

KELLEY, Robert L.
Battling the Inland Sea: American Political Culture, Public Policy, and the Sacramento Valley, 1850-1986 (Berkeley, CA: University of California Press, 1989). (YC.1993.b.2397)
Gold vs. Grain: The Hydraulic Mining Controversy in California's Sacramento Valley (Glendale, CA: Arthur H. Clark, 1959).
(10099.cc.11)

LEAVITT, Judith Walzwer.
The Healthiest City: Milwaukee and the Politics of Health Reform (Princeton, NJ: Princeton University Press, 1982). (X.529/48982)

LEE, Lawrence B.
"Environmental Implications of Governmental Reclamation in California," *Agricultural History* 49 (Fall 1975):223-30.
(Ac.3510.a)

LEUCHTENBURG, William E.
Flood Control Politics: The Connecticut River Valley Problem, 1927-1950 (Cambridge, MA: Harvard University Press, 1953). (08777.f.4)

MEINIG, Donald W.
The Great Columbian Plain: A Historical Geography, 1805-1910 (Seattle, WA: University of Washington Press, 1968). (X.800/3582)
The Shaping of Anerica (New Haven, CT: Yale University Press). Volume 1, *Atlantic America, 1492-1800* (1986) (YH.1987.b.367); Volume 2, Continental America, 1800-1867 (1993) (YC.1993.b.4883).
Southwest: Three Peoples in Geographical Change, 1600-1970 (New York: Oxford University Press, 1971). (X.800/5942)

PRESTON, William L.
Vanishing Landscape: Land and Life in the Tulare Lake Basin (Berkeley, CA: University of California Press, 1981). (X.805/2384)

SCARPINO, Philip V.
Great River: An Environmental History of the Upper Mississippi, 1890-1950 (Columbia, MO: University of Missouri Press, 1985). (YC.1988.a.3496)

SEARS, Paul B.
Deserts on the March (Norman, OK: University of Oklahoma Press, 1935). (07108.aa.24)

SHEROW, James E.
Watering the Valley: Development Along the High Plains Arkansas River, 1870-1950 (Lawrence, KS: University Press Of Kansas, 1990). (YA.1993.b.5055)

SMITH, Michael L.
Pacific Visions: California Scientists and the Environment, 1850-1915 (New Haven, CT: Yale University Press, 1987).

(YK.1988.b.5862)

STRONG, Douglas H.
Tahoe: An Environmental History (Lincoln, NE: University of Nebraska Press, 1984). (X.950/39125)

TERRIE, Philip G.
Forever Wild: Environmental Aesthetics and the Adirondack Forest Preserve (Philadelphia, PA: Temple University Press, 1985).

TOOLE, K. Ross.
The Rape of the Great Plains: Northwest America, Cattle, and Coal (Boston, MA, 1976).

WEBB, Walter Prescott.
The Great Frontier (Boston, MA: Houghton Mifflin, 1952). (X.700/1710)
The Great Plains (Boston, MA: Ginn, 1931). (010410.i.33)

WEIGOLD, Marilyn E.
The American Mediterranean: An Environmental, Economic and Social History of Long Island Sound (Lincoln, NE: University of Nebraska Press, 1979).

BIOGRAPHICAL STUDIES [See also Attitudes Toward Nature]

BONTA, Marcia Myers.
Women in the Field; America's Pioneering Women Naturalists (College Station, TX : Texas A & M University Press, 1991).

BRINK, Wellington.
Big Hugh [Hugh Hammond Bennett]: The Father of Soil Conservation (New York: Macmillan, 1951).

BROWER, David R.
For Earth's Sake: The Life and Times of David Brower (Salt Lake City, UT: Peregrine Smith Books, 1990).

CLEPPER, Henry.
Leaders in American Conservation. (New York, 1971)

COHEN, Michael P.
The Pathless Way: John Muir and the American Wilderness (Madison, WI: University of Wisconsin Press, 1984). (YA.1989.b.4272)

COOLEY, Richard A.
Alaska: A Challenge in Conservation (Madison, WI: University of Wisconsin Press, 1966). (X.800/2151)
    Politics and Conservation: The Decline of the Alaska Salmon (New York: Harper and Row, 1963). (7116.dd.24)

CUTRIGHT, Paul R.
Lewis and Clark: Pioneering Naturalists (Urbana, IL: University of Illinois Press, 1969). (X.311/2062)
    Theodore Roosevelt: The Making of a Conservationist (Urbana, IL: University of Illinois Press, 1985). (YC.1987.b.3915)
    Theodore Roosevelt, The Naturalist (New York: Harper, 1956). (10889.g.40)

DARRAH, William C.
Powell of the Colorado (Princeton, NJ: Princeton University Press, 1951). (10889.g.24)

FAUSOLD, Martin L.
Gifford Pinchot, Bull Moose Progressive (Syracuse, NY: Syracuse University Press, 1961). (X.709/873)

FICKEN, Robert E.
"Gifford Pinchot Men: Pacific Northwest Lumbermen and the Conservation Movement, 1902-1910," *Western Historical Quarterly* 13 (April 1982):165-78. (P.701/404)

FRADKIN, Philip L.
Wanderings of an Environmental Journalist in Alaska and the American West (Albuquerque, NM: University of New Mexico Press, 1993). (YA.1993.b.9965)

FRICK, George F. and Raymond P. Stearns.
Mark Catesby: The Colonial Audubon (Urbana, IL: University of Illinois Press, 1961). (10865.i.5)

GARTNER, Carol B.
Rachel Carson (New York: Frederick Ungar Publishing, 1983). (X.958/25688)

GLOVER, James M.
A Wilderness Original: The Life of Bob Marshall (Seattle, WA: The Mountaineers, 1986).

KELLEY, Elizabeth B.
John Burroughs: Naturalist (New York: Exposition Press, 1959). (10667.k.10)

LABASTILLE, Anne.
Women and Wilderness (San Francisco, CA: Sierra Club Books, 1980).

LENDT, David L.
Ding:The Life of Jay Norwood Darling (Ames, IA: Iowa State University Press, 1979).

LOWENTHAL, David.
George Perkins Marsh:Versatile Vermonter (New York: Columbia University Press, 1958). (10892.h.28)

LYON, Thomas J.
John Muir,Western Writers Series, No. 3 (Boise, ID: Boise State College, 1972). (X.0909/731)

MCCAY, Mary A.
Rachel Carson (New York: Twayne, 1993). (YA.1993.a.16803)

MACGEARY, Martin N.
Gifford Pinchot: Forester-Politician (Princeton, NJ: Princeton University Press, 1960). (10800.e.27)

MARGOLIS, John D.
Joseph Wood Krutch:A Writer's Life (Knoxville, TN: University of Tennessee Press, 1980). (X.800/34429)

MEINE, Curt.
Aldo Leopold: His Life and Work (Madison, WI: University of Wisconsin Press, 1987). (YC.1993.a.3273)

MORGAN, George T., Jr.
William B. Greeley, a Practical Forester, 1879-1955 (St. Paul, MN: Forest History Society, 1961).

NORWOOD, Vera.
"The Nature of Knowing: Rachel Carson and the American Environment," *Signs: Journal of Women in Culture and Society* 12 (Summer 1987):740-60. (P.521/2035)

PAUL, Sherman.
The Shores of America:Thoreau's Inward Vision
(Urbana, IL: University of Ilinois Press, 1959). (X.908/25039)

PINCHOT, Gifford.
Breaking New Ground (New York: Harcourt, Brace, 1947). [Autobiography]
The Fight for Conservation (London: Hodden & Stoughton, 1910). (8175.de.33) [Reprinted with introduction by Gerald D Nash: Seattle, WA: University of Washington Press, 1947) (X.519/5549)]

PINKETT, Harold T.
Gifford Pinchot: Private and Public Forester (Urbana, IL: University of Illinois Press, 1970).

RICHARDSON, Robert D.
Henry Thoreau:A Life of the Mind (Berkeley, CA: University of California Press, 1986). (YC.1988.b.465)

ROBERTSON, Janet.
Those Magnificent Mountain Women:Adventures in The Colorado Rockies (Lincoln, NE: University of Nebraska Press, 1990).

RODGERS, Andrew Penny.
Bernard Edward Fernow:A Story of North American Forestry (Princeton, NJ: Princeton University Press, 1951). (10891.c.31)

RONALD, Ann.
The New West of Edward Abbey (Albuquerque, NM: University of New Mexico Press, 1982). (X.800/40696)

SHANKLAND, Robert.
Steve Mather of the National Parks (New York: Knopf, 2nd ed. rev., 1954).

SMITH, Herbert F.
John Muir, Twayne United States' Author Series (New York: Twayne Publishers, 1965). (X.909/9776)

STERLING, Philip.
Sea and Earth: The Life of Rachel Carson (New York: Thomas Y. Crowell, 1970).

SWAIN, Donald C.
Wilderness Defender: Horace M. Albright and Conservation (Chicago IL: University of Chicago Press, 1970) (X.329/1118)

TANNER, Thomas, ed.
Aldo Leopold: The Man and His Legacy (Ankeny, IA: Soil Conservation Society, 1987).

TEALE, Edwin Way.
The Wilderness World of John Muir (Boston, MA: Houghton Mifflin, 1982).

WADLAND, John Henry.
Ernest Thompson Seton: Man in Nature and the Progressive Era, 1880-1915 (New York: Arno Press, 1978).

WHITE, Graham and John Maze.
Harold Ickes of the New Deal: His Private Life and Public Career (Cambridge, MA: Harvard University press, 1985). (YC.1988.b.6096)

WILD, Peter.
Barry Lopez, Western Writers Series (Boise, ID: Bosie State University, 1984). (X.0909/731)

WOLFE, Linnie Marsh.
Son of the Wilderness: The Life of John Muir (New York: Knopf, 1945). (Mic.A.8939)

ZUCKER, Norman L.
George W. Norris, Gentle Knight of American Democracy (Urbana, IL: University of Illinois Press, 1966). (X.700/1914)

ENVIRONMENTAL ETHICS

ALTFIELD, Robin.
The Ethics of Environmental Concern (New York: Columbia University Press, 1983).

BARBOUR, Ian G.
Earth Might Be Fair: Reflections on Ethics, Religion and Ecology (Englewood Cliffs, NJ: Printice-Hall, 1972). (X.100/10442)
    Technology, Environment and Human Values (New York: Praeger, 1980). (X.622/18997)
    Western Man and Environmental Ethics, Attitudes toward Nature and Technology (Menlo Park, CA: Addison Wesley, 1973). (X.319/6381)

BLACKSTONE, William T., ed.
Philosophy and the Environmental Crisis (Athens, GA; University of Georgia Press, 1974).

BOOKCHIN, Murray.
The Ecology of Freedom: The Emergence and Dissolution of Hierarchy (Palo Alto, CA: Cheshire Books, 1982). (X.529/52873)
    The Philosophy of Social Ecology: Essays on Dialectial Naturalism (Montreal: Black Rose Books, 1990). (YA.1992.a.18623)
    Remaking Society: Pathways to a Green Future (Boston, MA: South End Press, 1990). (YA.1991.a.19227)
    Toward an Ecological Society (Montreal: Black Rose, 1980).

BORMAN, F. Herbert and Stephen R. Kellert, eds.
Ecology, Economics, Ethics: The Broken Circle (New Haven, CT: Yale University Press, 1991). (YC.1992.b.533)

BORRELLI, P., ed.
Crossroads: Environmental Priorities for the Future (Washington, D.C.: Island Press, 1988).

CAHN, Robert.
Footprints on the Planet: A Search for An Environmental Ethic (New York: Universe, 1978).

CALLICOTT, J. Baird.
In Defense of the Land Ethic: Essays in Environmental Philosophy (Albany, NY: State University of New York Press, 1989). (YH.1990.b.203)

DEVALL, Bill.
Simple in Means, Rich in Ends: Practicing Deep Ecology (London: Green Print, 1990). (YK.1991.a.7631)

DEVALL, Bill and George Sessions.
Deep Ecology: Living as if Nature Mattered (Salt Lake City, UT: Peregrine Smith Books, 1985).

DOUGLAS, Mary and Aaron Wildavsky.
Risk and Culture: An Essay on the Selection of Technological and Environmental Dangers (Berkeley, CA: University of California Press, 1983). (X.629/18726)

EHRLICH, Paul Ralph, and Anne H. Ehrlich.
Population, Resources, Environment: Issues in Human Ecology (San Francisco, CA: W.H. Freeman and Co., 2nd ed., 1972). (W.P. 58/52)

EHRLICH, Paul R., et al.
Human Ecology: Problems and Solutions (San Francisco, CA: W.H. Freeman and Co., 1973). (X.311/3412)

FOX, Michael Allen and Leo Groarke, eds.
Nuclear War: Philosophical Perspectives (New York: Lang, 1985).

HARGROVE, Eugene.
Foundations of Environmental Ethics (Englewood Cliffs, NJ: Prentice-Hall, 1988). (YH.1989.a.833)
    ed. *Beyond Spaceship Earth: Environmental Ethics and the Solar System* (San Francisco, CA: Sierra Club, 1986).
    ed. *Religion and Environmental Crisis* (Athens, GA: University of Georgia Press, 1986). (YH.1989.a.75)

HORKEIMER, Max.
The Eclipse of Reason (New York: Plenum, 1974).

KATZ, Eric.
"A Selective Bibliography of Environmental Ethics, 1983-1987," *Research in Philosophy and Theology* 9 (1988).

LESTER, James P., ed.
Environmental Politics and Policy, Theories and Evidence (Durham, NC: Duke University Press, 1989). (YC.1993.b.2128)

NAESS, Arne.
Ecology, Community and Lifestyle, Outline of an Ecosophy (Cambridge: Cambridge University Press, 1989). (YC.1989.b.2954)
    "The Shallow and the Deep, Long-Range Ecology Movements," *Inquiry* 16 (1973):95-100.
    Spinoza and the Deep Ecology Movement (Delft: Eburon, 1993). (Ac.2720 [vol. 67])

NASH, Roderick.
The Rights of Nature: A History of Environmental Ethics (Madison, WI: University of Wisconsin Press, 1989). (YC.1991.a.3203)

NICHOLSON, Edward Max.
The Environmental Revolution: A Guide for the New Masters of the World (London: Hodder and Stoughton, 1970). (X.329/3997)
    The New Environmental Age (Cambridge:

Cambridge University Press, 1987).
(YC.1987.b.5685)

NORTON, Byron G.
Why Preserve Natural Variety? (Princeton, NJ: Princeton University Press, 1988), (YH.1989.a.910)

OATES, D.
Earth Rising: Ecological Belief in an Age of Science (Corvallis, OR: Oregon State University Press, 1989).

PASSMORE, John.
Responsibility for Nature: Ecological Problems and Western Traditions (London: Duckworth, 1974). (X.320/4596)

PERROW, Charles.
Normal Accidents: Living with High Risk Technologies (New York: Basic Books, 1984). (X.622/25957)

PIRAGES, Dennis Clark, and Paul Ralph Ehrlich.
Ark II: Social Responses to Environmental Imperatives (San Francisco, CA: W.H. Freeman, 1974). (X.510/8415)

ROLSTON, Holmes.
Environmental Ethics: Duties to and Value in the Natural World (Philadelphia, PA: Temple University Press, 1988). (YA.1990.b.2635)

ROSZAK, Theodore.
Person-Planet: The Creative Disintegration of Industrial Society (London: Gollancz, 1979). (X.529/34422)
    Where the Wasteland Ends: Politics and Transendence in a Postindustrial Society (London: Faber, 1973). (X.529/15892)

SANTMIRE, H, Paul.
The Travail of Nature: The Ambiguous Ecological Promise of Christian Theology (Philadelphia, PA: 1985)

SCHERER, Donald, ed.
Upstream/Downstream: Issues in Environmental Ethics (Philadelphia, PA: Temple University Press, 1990). (YA.1993.a.17631)

SCHERER, Donald and Thomas Attig, eds.
Ethics and the Environment (Englewood Cliffs, NJ: Prentice-Hall, 1983). (X.529/56360)

SCHUMACKHER, Ernest F.
Small is Beautiful: A Study of Economics as if People Mattered (London: Blond and Briggs, 1973). (X.529.16323)

SIRY, Joseph V.
Marshes of the Ocean Shore: Development of an Ecological Ethic (College Station, TX: Texas A&M Press, 1984).

STEWART, Claude Y., Jr.
Nature in Grace: A Study in the Theology of Nature (Macon, GA: Mercer University Press, 1983).

TOBIAS, Michael, ed.
Deep Ecology (San Diego, CA: Avant Books, 1988). (YA.1990.a.16274)

COMPARATIVE STUDIES
(Selected)

BAHRO, Rudolf.
Building the Green Movement (translated by Mary Tyler (London: GMP, 1986). (YC.1987.a.1299)

BAILES, Kendall E., ed.
Environmental History: Critical Issues in Comparative Perspective (Lanham, MD: University Press of America, 1985).

BENNETT, John.
The Ecological Transition: Cultural Anthropology and Human Adaptation (New York: Pergamon, 1976). (X.319/16356)

BOWLER, Peter J.
The Fontana History of the Environmental Sciences (London: Fontana, 1992). (YK.1992.a.10377)

BRAMWELL, Ann.
Ecology in the 20th Century: A History (New Haven, CT: Yale University Press, 1989). (YC.1991.b.3367)

BRIMBLECOMBE, Peter, and Christian Pfister, eds.,
The Silent Countdown: Essays in European Environmental History (Berlin: Springer-Verlag, 1990).

BROWN, Michael H. and John May.
The Greenpeace Story (London: Dorling Kindersley, 1991). (YK.1992.b.1886)

CAHN, Robert and P. Cahn.
"Did Earth Day [22 April 1970] Change the World?" *Environment* (September 1990):16-20, 36-43.

CALLICOTT, J. Baird and Roger T. Ames, eds.
Nature in Asian Traditions of Thought: Essays in Environmental Philosophy (Albany, NY: State Univesity of New York Press, 1989). (YH.1990.b.196)

CAPRA, Fritjof, and Charlene Spretnak in collaboration with Rudiger Lutz,
Green Politics (London: Hutchinson, 1984). (X.809/64382)

COMMONER, Barry,
The Closing Circle: Confronting the Environmental Crisis (London: Cape, 1972). (X.329.5338)
    Making Peace with the Planet (New York: New Press, 1992). (YC.1993.a.2408)
    The Poverty of Power: Energy and the Economic Crisis (London: Cape, 1976). (X.529/31243)

DAY, David.
The Eco Wars: A Laymans's Guide to the Ecology Movement (London: Paladin, 1991). (YC.1991.a.5192)
    The Environmental Wars: Reports from the Front Lines (New York: St. Martin's Press, 1989). (YA.1991.a.17737)
    The Whale War (London: Routledge and Kegan Paul, 1987). (YC.1987.b.3313)

DOMINICK, Raymond H.
The Environmental Movement in Germany: Prophets and Pioneers, 1871-1971 (Bloomington, IN: Indiana University Press, 1992). (YA.1993.b.6413)

FRANKE, Richard W. and Barbara H. Chasin,
Seeds of Famine: Ecological Destruction and the Development Dilemma in the West African Sahel (Montelair, NJ: Allenheld, Osmun, 1980). (X.322/11754)

HUNTER, Robert.
The Greenpeace Chronicle (London: Pan Books, 1980). [Originally published in New

York as *Warriors of the Rainbow: A Chronicle of the Greenpeace Movement*] (X.908/43076)

HURLEY, Patrick.
The Greenpeace Effect (London: Macmillan, 1991). (YK.1991.b.7781)

KELLEY, Donald R., ed.
The Energy Crisis and the Environment: An International Perspective (New York: Praeger, 1977). (X.320/11039 Woolwich)

LANGGUTH, Gerd.
The Green Factor in German Politics: From Protest Movement to Political Party (Boulder, CO: 1986). [German Edition: Zurich: Edition Interfrom, 1984, (YA.1988.a.5783)]

MCCORMICK, John.
Acid Earth: The Global Threat of Acid Pollution (London: Earthscan, 2nd ed., 1989). (YK.1990.a.5209)
The Global Environment Movement: Reclaiming Paradise (London: Belhaven Press, 1989). (YC.1991.b.4361)

MCNEILL, William H.
The Human Condition: An Ecological and Historical View (Princeton, NJ: Princeton University Press, 1980). (X.529/37782)
Plagues and Peoples (Oxford, UK: Blackwell, 1977). (X.809/41966)

MANES, C.
Green Rage: Radical Environmentalism and the Unmaking of Civilization (Boston, MA: Little, Brown, 1990).

MANNION, A. M.
Global Environmental Change: A Natural and Cultural Environmental History (Harlow: Longman Scientific and Technical, 1991). (YC.1991.b.1155)

PAPADAKIS, Elim.
The Green Movement in Germany (London: Croom Helm, 1984). (X.809/60472)

PEPPER, David, et al.
The Roots of Modern Environmentalism (London: Routledge, 1989), (YC.1989.a.5332)

PILAT, Joseph F.
Ecological Politics: The Rise of the Green Movement (Beverly Hills, CA: Sage Center for Strategic and International Studies, 1980). (X.0709/603(77))

POWELL, Joseph Michael.
Environmental Management in Australia, 1788-1914: Guardians, Improvers and Profit, An Introductory Survey (Melbourne: Oxford University Press, 1976). (X.320/5129)

SIMMONS, Ian G.
Environmental History, A Concise Introduction (Oxford: Blackwell Publishers, 1993). (YC.1993.b.8944)

THOMAS, William L., Jr., ed.
Man's Role in Changing the Face of the Earth, 2 vols. (Chicago IL: University of Chicago Press, 1956).

TUCKER, Richard P. and J. F. Richards, eds.
Global Deforestation and the Nineteenth-Century World Economy (Durham, NC: Duke University Press, 1983). (X.520/34456)

WALL, Derek.
Green History: A Reader in Environmental Literature, Philosophy and Politics (London: Routledge, 1994). (YC.1994.b.528)

WATKINS, T.H.
Righteous Pilgrim: The Life and Times of Harold L. Ickes, 1874-1952 (New York: Henry Holt,

1990). (YA.1993.b.4598)

WORSTER, Donald. ed.
"World Without Borders: The Internationalizing of Environmental History," in *Environmental History: Critical Issues in Comparative Perspective* Kendall Bailes, ed. (Lanham, MD: University Press of America, 1985).

A CHRONOLOGY of MAJOR LEGISLATION and EVENTS in AMERICAN ENVIRONMENTAL HISTORY

1849	Department of the Interior established
1864	Yosemite Valley, CA, reserved as a state park
1871	U.S. Fish Commission created
1872	Yellowstone National Park established
1875	American Forestry Association organized
1877	Desert Land Act
1879	U.S. Geological Survey established
1881	Division of Forestry created in Department of Agriculture
1885	Adirondack Forest Preserve established by New York
1886	New York Audubon Society organized
1891	Forest Reserves Act
1891	Yosemite National Park established
1892	Sierra Club founded
1894	Carey Act
1897	Forest Management Act
1899	River and Harbor Act
1900	Lacey Act
1902	Reclamation (Newlands) Act creates Reclamation Service [in 1923 renamed the Bureau of Reclamation] in the Department of the Interior; begins federal reclamation program
1904	Florida's Pelican Island became the first federal wildlife reservation
1905	National Audubon Society organized
1906	Antiquities Act
1910	Ballinger-Pinchot controversy
1913	Migratory Bird Act
1913	Hetch Hecty Valley Reservoir controversey concluded
1916	National Park Service Act
1918	Migratory Bird Treaty Act
1918	Save-the-Redwoods League founded
1920	Mineral Leasing Act
1920	Fedeeral Water Power Act
1922	Colorado River Compact

1924	Oil Pollution Control Act
1924	Teapot Dome scandal
1924	Clarke-McNary Act
1928	McSweeney-McNary Act
1928	Congresses authorizes construction of Boulder [Hoover] Dam
1933	Civilian Conservation Corps established
1933	Tennessee Valley Authority created
1934	Taylor Grazing Act
1935	Soil Conservation Act
1936	National Wildlife Federation founded
1937	Federal Aid in Wildlife Restoration Act
1937	Bonneville Dam completed
1940	U.S. Fish and Wildlife Service created
1941	Grand Coulee Dam completed
1946	U.S. Bureau of Land Management established
1948	Federal Water Pollution Control Law enacted
1949	National Trust for Historic Preservation charted by Congress
1956	Echo Park Dam project in Dinosaur National Monument defeated
1956	Water Pollution Control Act
1960	Multiple Use-Sustained Yield Act
1962	White House Conference on Conservation
1963	Clean Air Act
1964	Wilderness Act
1965	Land and Water Conservation Fund Act
1965	Storm King case [Hudson River]
1966	National Historical Preservation Act
1966	Endangered Species Act
1968	National Wild and Scenic Rivers Act
1968	National Trails System Act
1968	Redwoods National Park established
1969	Santa Barbara, CA oil spill
1969	Friends of the Earth founded
1969	Greenpeace organized
1970	National Environmental Policy Act

1970	Solid Waste Disposal Act
1970	Clean Air Act [strengthened 1963 act]
1970	First "Earth Day" [April 22]
1970	National Oceanic and Atomospheric Administration created
1970	Environmental Protection Agency created
1972	Federal Water Pollution Control Act
1972	Federal Environmental Pesticide Control Act
1972	Ocean Dumping Act
1972	Coastal Zone Management Act
1973	Endangered Species Act [expansion of 1966 Act]
1974	Safe Drinking Water Act
1976	Federal Land Policy and Management Act
1976	Resource Conservation and Recovery Act
1976	Toxic Substances Control Act
1977	Clean Air Act [amendments]
1977	Federal Water Pollution Control Act [amendments]
1977	Surface Mining Control and Reclamation Act
1978	National Energy Act
1980	Alaska National Interest Lands Conservation Act
1980	Fish and Wildlife Conservation Act
1981	Earth First! organized
1989	Oil spill in Prince William Sound, Alaska